Power and the Public Interest

POWER
AND THE
PUBLIC INTEREST

The Memoirs of Joseph C. Swidler

Foreword by Lee White

Edited and Annotated by A. Scott Henderson

The University of Tennessee Press / Knoxville

First Edition.

This book is printed on acid-free paper.

Library of Congress Cataloging in Publication Data

Swidler, Joseph C. (Joseph Charles), 1907–
Power and the public interest : the memoirs of Joseph C. Swidler / edited and annotated by A. Scott Henderson; foreword by Lee White.
p. cm.
Includes bibliographical references and index.
ISBN 1-57233-183-6 (cl.: alk paper)
1. Swidler, Joseph C. (Joseph Charles), 1907–
2. Energy policy—United States—History.
3. Power resources—Government policy—United States—History.
4. Tennessee Valley Authority—Officials and employees—Biography.
5. United States. Federal Power Commission—Officials and employees—Biography.
6. New York (State). Public Service Commission—Officials and employees—Biography.
I. Henderson, A. Scott.
II. Title.

HD9502.U52 S95 2002
333.79'092—dc21 2001006277

To Gretchen Swidler

Contents

PART VI. Chairman of the Federal Power Commission, 1961–1965

PART VII. Back to Private Practice, 1966–1970

PART VIII. Chairman of the New York State Public Service Commission, 1970–1974

PART IX. Final Years, 1974–1997

Illustrations

Following Page 116

Foreword

In 1950, when I arrived at the Tennessee Valley Authority as a young lawyer, Joe Swidler was ensconced in the position of general counsel. He was more than that, though. He had the agency so sensitized to the dominant role he played that just about everyone checked with the legal division even before going to the restroom. After only a couple of months at TVA, I went to an internal meeting on a policy issue. When I told Joe what had transpired and why I thought it was wrong, he put in a joint call to the heads of the division of reservoir properties and the division of water control. He proceeded to tell them that they had reached the wrong decision. When they replied that he had had one of his lawyers present, and that I, the lawyer, should have spoken up, Joe said: "He is new here and assumed that you two division chiefs knew what you were talking about; he won't make that mistake again."

Most people, including high-ranking government officials, go through life unwilling to rock the boat. Consequently, they only have modest achievements to show for their efforts. That was neither the substance nor the style of Joe Swidler. Admittedly, Joe could be brusque, and I do not think he ever won a prize for being tactful. He was less concerned, however, with what others thought of him than with getting things done and implementing various reforms.

The TVA was not the only organization where Joe left his stamp. When he went to Washington at the beginning of the Kennedy administration to head the Federal Power Commission, he found an agency described in a report to the new president as the worst of all the regulatory bodies. Some political appointees can keep a good organization rolling along nicely, but it is a rare individual who can take an ineffective agency with very low morale and convert it into an efficient and greatly respected one. By any measure, Joe was instrumental in achieving the FPC's impressive turnaround. Of course, he had some talented helpers and some highly capable commission members. His take-charge style was not free of controversy, and his run-in with a fellow commissioner was embarrassing and disruptive. Joe was accused of not being sufficiently sympathetic to consumers, and while it is true that he did not always side with groups that were his natural allies, even those who disagreed with him did not challenge his integrity.

One of Joe's most enduring qualities was his tendency and ability to negotiate. My first experience with this characteristic was when he offered me the TVA job. I told him I was committed to a thirty-day stint of active duty as a reserve officer at Fort Monmouth, New Jersey. When he told me to get out of it, I said I did not think I could or should. His response was, "You better hope I have not filled the spot." Another example came during a tour of the Soviet Union's hydroelectric facilities in 1962. The delegation, headed by U.S. Interior Secretary Stewart Udall, was told in Moscow that we could take all the photographs we wanted. When we arrived at the Bratsk Dam in Siberia, the project manager said that no pictures could be taken. Udall explained what we had been told in Moscow. The manager was not impressed and said he was in charge at the dam site. Udall protested a bit, then decided that it was time to yield the point. But not Joe. He took over and gave the project manager such a hard time he gave in. The best part of the story is that Joe had run out of film; he had to pretend he was taking pictures.

In 1969 New York Governor Nelson Rockefeller sent an emissary to sound out Joe about taking over the chairmanship of the New York State Public Service Commission. It was deemed to be one of the worst in the country and had the reputation of being a resting place for political hacks. Joe asked me what I thought about his taking over a state commission after he had been chairman of a federal agency. I told him that his reputation was so strong that it would automatically enhance the importance of the assignment simply because he was willing to accept it. With a typical Swidler quip, he responded: "I guess you want to take over my clients." He took the position and, as with the FPC, turned that agency around. When Joe returned to Washington after his days in New York came to a close, we had lunch again to discuss what he should do next. He had already accepted a position as the president of an engineering consulting firm. I told him he was a lawyer and should practice rather than embark on a course of hustling utility executives for a consulting firm. And this time there were no clients for me to inherit.

Joe Swidler led a long, interesting, and productive life. He used his sharp intellect and his get-it-done style to accomplish a great deal. His memoir is not only a fascinating retelling of his involvement in the electric power and natural gas industries, but also an important recounting of significant political and policy developments between 1933 and the early 1970s. This story will be of interest to many different readers, especially those who wish to serve the public interest.

Lee White
Washington, D.C.

Acknowledgments

I owe thanks to many people for helping me edit and annotate this memoir. Jack Bizot, Kurt Blocher, Mark Cenczyk, Wayne Millan, John Poling, Terrianne Schulte, and Peggy Tuck Sinko provided excellent research assistance; regrettably, this space does not allow me to acknowledge their contributions adequately. The librarians at Furman University, especially Brenda Custard, Mary Fairbairn, James Harper, John Payne, and Steve Richardson, cheerfully came to my rescue whenever I had a question or needed the services of interlibrary loan. Elsewhere, librarians at Clemson University, Indiana University, the University of Memphis, the TVA Corporate Library, and the University of South Carolina assisted me in finding various items in their collections.

The alumni offices of several universities were also helpful. They included Albany Law School, Amherst College, Boston University, Clark University, Colgate University, Columbia University, Cornell University, Georgetown University, Harvard University, Northwestern University, Pennsylvania State University, Princeton University, Stanford University, United States Military Academy, University of Chicago, University of Florida, University of Minnesota, University of Southern California, University of Washington, University of Wisconsin, and Yale University.

I called upon many other organizations and agencies to verify and clarify certain parts of Swidler's narrative. Among them were the American Public Power Association; Boston Historical Society; Brownsville (Tennessee) Utility Department; California State Bar Association; Carnegie Library of Pittsburgh; Chattanooga Electric Power Board; Chattanooga–Hamilton County Public Library; Chicago Historical Society; Chicago Public Library; Columbia Oral History Research Office; District of Columbia Bar Association; Edison Electrical Institute; Electric Power Research Institute; Federal Energy Regulatory Commission; Federal Trade Commission; Historical Society of Washington, D.C.; Library of Congress; Linda Hall Library of Science, Engineering, and Technology; National Archives and Research Administration; Naval Facilities Engineering Command; Naval Historical Center; New York City Bar Association; Oregon Democratic Party; Paducah (Kentucky) Electric Power; Pittsburgh Historical Society; Rhode Island State Bar Association; Smithsonian National Portrait Gallery; Tennessee Valley Authority; United States Central Intelligence Agency; Washington, D.C., Public Library.

Scholars, government officials, colleagues, and friends who provided comments, suggestions, and support were Ed Best, Amy Blackwell, Chris Blackwell, Mike Bressler, Deborah Cherry, Jan Comfort, Rose Crisostomo, Jon Dembo,

Erwin C. Hargrove, John Haynes, Nelly Hecker, Brad Hunt, Gail Jenkinson, Hugh Jenkinson, Alvin Kaufman, Charles J. McCarthy, Harold Platt, Sue Presnell, Nancy Proctor, Lesley Quast, Patrick Quinn, Sarah Robinson, Mark Rose, Adam Schwartz, Amanda Seligman, Denise Smith, Julie Thomas, Joseph Volpe Jr., David Warrington, and Judith Wright. I also appreciate the financial support of the Joseph C. Swidler estate and a Research and Professional Development Grant from Furman University.

Several individuals deserve special recognition. Louis Blair's friendship and support have gotten me through many low moments. Jeanne Clarke, Steven M. Neuse, and Michael E. Parrish reviewed the unedited memoir and made excellent suggestions. Lee White, a former adviser to two presidents and one of Swidler's closest friends, kindly agreed to write the foreword. Patricia B. Ezzell, historian of the TVA, was always ready and willing to help me with a research request. Ed Frank assisted me in obtaining several important oral history transcripts from the Mississippi Valley Collection at the University of Memphis. Brad Gernand generously guided me through the Swidler Papers and several other collections at the Library of Congress. At the National Archives and Records Administration in East Point, Georgia, Arlene Royer helped me track down items from the hundreds of boxes containing TVA material. Alan Lessoff provided crucial advice at several junctures; I am indebted to his keen sense of both the big and the small historical picture. Gail Radford read the unedited memoir and offered valuable suggestions and comments during the entire editing process. Her personal commitment to the project, her knowledge of New Deal America, and her insightful feedback were indispensable. The works of Jon Dembo provided standards of editorial scholarship that I tried to emulate. I would also like to thank, at the University of Tennessee Press, Joyce Harrison for her patience and counsel, as well as June Hussey, Stan Ivester, and Joanna Juzwik McDonald for their fine editorial assistance. I cannot thank Richard Prior enough for his support and forbearance. Of course, I take full responsibility for any and all errors associated with the editing and annotation of this memoir.

Special recognition needs to be extended to three other individuals. Ann and Mark Swidler enthusiastically supported this project from the very beginning, providing candid appraisals of their father and his colleagues. Gretchen Swidler's recollections of her husband and their life together were also invaluable. I benefited in numerous ways from her gracious encouragement and assistance. By sharing their memories with me, the family of Joe Swidler gave me a better understanding of his life, his personality, and his accomplishments.

Editor's Introduction

In early May 1997, obituaries across the country announced that Joseph Swidler had died at the age of ninety. Sixty-four years earlier, an enthusiastic Swidler had journeyed to Washington, D.C., to help implement New Deal economic reforms. What began as youthful inspiration over the vision of Franklin Roosevelt turned into a lifetime of public service. Indeed, Swidler belonged to a generation of talented professionals who devoted most of their energies to serving public, not private interests. They included, among others: Wilbur Cohen ("father" of Social Security), Harry Hopkins (director of the Federal Emergency Relief Administration), and Harold Ickes (head of the Public Works Administration). Swidler himself played an important part in developing the Tennessee Valley Authority (TVA) and in expanding federal and state utility regulation.

Neither Swidler nor his contemporaries became public servants or policymakers *ex nihil.* Like many others, Swidler was trained as a lawyer. This link between legal expertise and policymaking was logical since the language of the administrative state was, and continues to be, the language of the law. Being a lawyer, however, did not restrict Swidler's thinking or outlook. As his recollections attest, he always kept a larger context in mind. No matter how complicated the legalese or economic calculations might become, he never forgot they would affect real people. While this did not obviate the necessity of embracing the quotidian realities of policymaking, it insured that Swidler never became the type of insensate functionary whom unsympathetic commentators subsequently derided as nameless, faceless bureaucrats.

Swidler was also Jewish. This meant discrimination could and did limit his initial career opportunities. His first significant encounter with anti-Semitism was fateful. Applying to various law firms during the Depression, he found their doors politely, but firmly, shut. A gentleman's agreement still prevented Jews from being employed in many law offices. Swidler eventually gained a position working for David Lilienthal, a coreligionist who became an important New Deal official himself. It was Lilienthal who persuaded Swidler to begin his almost twenty-five-year association with the TVA. For Swidler, public sector employment offered advantages not found in the private sector. There were fewer discriminatory barriers, more chances to effect social reform, and greater opportunities to manifest one's patriotism and assimilation. Whether this last point was important for Swidler remains unclear; the first two certainly were.

Swidler's early experiences instilled in him a lifelong fascination with the production and distribution of energy. Had he lived fifty years earlier, this interest might have inclined him toward a scientific career. He was destined, though, to make a different set of contributions. In regard to power production, the challenge of the first half of the twentieth century was not so much technological as it was logistical and political. The logistical dimension involved the question of how to supply power (electric as well as natural gas) to the greatest number of people; the political embraced whether the federal and/or state governments should play a role in this process. From the New Deal forward, decision makers answered these and related questions in ways that required expansion of the national government, a process scholars call "state-building." Swidler's assessments demonstrate that state-building was frequently ad hoc, unfocused, and unplanned. Phrased differently, his reflections are not a predictable story of the predestined liberal state, but a tale of contingency, detours, and policy roads not taken.

Despite Swidler's commitment to serving others, he moved back and forth between the public and private sectors, using the knowledge gained in one to aid him in the other. By doing so, he was not engaging in what later critics would call "influence peddling." Painfully conscious of his actions—and how others might perceive them—Swidler maintained strict professional scruples. Given current cynicism over the behavior of public officials, Swidler's account is refreshing without being naive, instructive without being moralistic.

Placing Swidler in a particular political category is not easy. He was admittedly partisan—specifically a Democrat. Yet he was unapologetic about staking out his own philosophical terrain. For instance, his opinions on "sunshine laws" and the financing of public utilities broke from traditional liberal thinking. And while he made clear distinctions between public and private power (in both senses of the word "power"), his world was gray, not a Manichean black or white. He avoided the Progressive tendency to view policymaking as an unending struggle between the masses and big business. More often, he saw it as an arena in which one had to balance public concerns with competing, sometimes legitimate claims presented by corporate interests. As he would argue toward the end of his life, "The relations between government and industry are a complex problem and do not lend themselves to simple or formulaic solutions." Though Swidler usually wanted the scales tipped in favor of the average citizen, he remained a strong advocate of capitalism; markets, not the state, were destined to structure economic life. In this respect, Jordan Schwarz's general description of New Dealers can be applied to Swidler: "They were building capitalism with a human face. They would make capitalism work more fairly and equitably for middle-class investors, consumers, and workers." Swidler's views on the production and regulation of power were consistent with this emphasis. Only when private enterprise was unable or unwilling to accomplish certain objectives did he support direct government intervention.[1]

Swidler's memoir helps us better understand developments that continue to shape contemporary American life. Energy—its availability, generation, and distribution—has been a critical component of industrial and economic expansion. The 1973 oil boycott, rising energy costs, the Gulf War in 1991, and the California power shortages during 2000–2001 underscored this point. These events occurred against the steady drumbeat of deregulation, an ideological riptide that, beginning with the Public Utility Regulatory Policies Act in 1978, helped spawn a growing number of unregulated ("independent") power producers. By the early 1990s, the issue of utility deregulation had become the focus of many state and local referenda. Whether or not energy, like air travel and surface transportation, will become a predominantly deregulated industry remains uncertain. Based on a lifetime of experience, Swidler clearly thought knee-jerk deregulation possessed inherent disadvantages. His account clarifies how and why a previous arrangement known as the "regulatory compact" was so successful, and how it might be modified—instead of abandoned—to meet future needs.[2]

A few comments about the memoir's structure are necessary. Swidler worked on this project during the last ten years of his life, leaving a manuscript of approximately seven hundred pages. Starting with his birth and childhood, Swidler traced his life up through the early 1970s. Despite the memoir's length, Swidler wrote sparingly of his personal life and feelings. He says very little about Harry R. Booth, an individual with whom he worked closely, both under Lilienthal and in a law firm they started together. Or, to cite another example, Swidler notes his marriage to Janet Voss in 1933, yet mentions almost nothing else about their relationship (they were divorced in 1943). Had Swidler addressed these points, he might have given his narrative some added human interest value, but our understanding of his public duties or persona would probably not have been enhanced. In light of his comments concerning Lilienthal's *Journals,* Swidler obviously believed that autobiographical works could become self-serving, sometimes even annoyingly self-confessional. Swidler's memoir avoids these defects.[3]

While Swidler was a clear and logical writer, the original manuscript contained lengthy passages that were overly technical and redundant. I have omitted this material, combined some chapters, and rearranged others. I retained most of Swidler's subheadings and added a few others. I have also written headers for the memoir's nine sections and inserted footnotes to clarify factual points. Within the text, parenthetical material is Swidler's; bracketed material has been added. So that future scholars can compare the unedited and edited versions, copies of the former are located in the Library of Congress (home to Swidler's papers) and the University of Tennessee. Accompanying these copies are versions of the edited manuscript in which all excisions and rearrangements are identified.

Swidler initially wanted to entitle his memoir "Adventures in Public Administration." That title might not have been as contradictory as it sounds. Swidler was involved in two of America's most important federal agencies (the TVA and the Federal Power Commission) and the most significant state regulatory agency (the New York Public Service Commission). His career and associations were intertwined with key public figures, including David Lilienthal, Wendell Willkie, Harold Ickes, Dean Acheson, Abe Fortas, Arthur Goldberg, Nelson Rockefeller, Strom Thurmond, John F. Kennedy, Henry "Scoop" Jackson, Lyndon Johnson, and scores of elected officials. He also witnessed most of the signal events of the modern age: the Depression, World War II, the Cold War, and America's emergence as the sole superpower. His duties took him to remote regions of rural America, vast power complexes in Siberia, Carpenters Hall in London, and Lyndon Johnson's ranch. He helped craft and administer the nation's energy policies during times of peace as well as economic instability. His career, spanning more than six decades, was truly an "adventure."

PART I

Education, Private Practice, and the Pull of Public Service, 1907–1933

The world of Swidler's youth would have been familiar to many American immigrants. His parents, along with countless other Jews, came to the United States seeking sanctuary from the state-sanctioned repression of late nineteenth-century Russia. Unlike many of their compatriots, the Swidlers were not penniless when they arrived. They joined a network of relatives in Chicago, two of whom—a maternal aunt and uncle—were already "rich," according to Swidler. Swidler's father worked as a tailor, then became a partner in a soft-drink bottling firm. These advantages notwithstanding, Swidler was probably accurate when he stated that the family "lived on the margin of economic security." The situation became more precarious when his father's investment in a Miami restaurant failed. Only through part-time work and the generosity of a better-off relative was Swidler able to receive a formal education beyond high school.

As far as his family was concerned, the verbal skills Swidler displayed during childhood sealed his vocational fate: He would become a lawyer. Though he expresses some regret over this foreordained conclusion, he seemingly had no second thoughts at the time. To the contrary, his memoir indicates that be began following Supreme Court decisions even before he graduated from high school. The justices he liked most, all advocates of what eventually became known as "sociological jurisprudence," were an indication of his evolving political sensibilities (sociological jurisprudence often dictated a more active role for courts). Despite his judicial preferences, Swidler labeled himself a "fiscal conservative." Yet the rest of his memoir illustrates that Swidler's support of fiscal prudence did not make him an advocate of minimal government; he simply wanted one that could pay for the ambitious programs it undertook.

Swidler's post-secondary education was geographically and philosophically peripatetic: He attended four colleges in as many years. He seems to have hit his intellectual stride at the University of Chicago Law School where he was challenged, if not always stimulated, by its well-known faculty. In an age when most law schools required prospective students to possess a bachelor's degree, Swidler, having completed only five semesters of college work, was fortunate that Chicago retained flexible admissions criteria.

Through David Lilienthal, one of the country's rising public figures, Swidler was introduced to the field of utility regulation. Knowledge about the utility industry is necessary to understand why this form of regulation was important. Nineteenth-century capital constraints prevented most companies from engaging in, much less profiting from, energy production. Those that did organized themselves into monopolies called "trusts." After trusts were outlawed in the late nineteenth century, entrepreneurs created holding companies—companies that held controlling stock in *other* corporations. In the utility field, these other corporations were usually so-called operating companies, the entities that actually produced energy. While holding companies initially brought much-needed capital and rationalization to the utility industry, they were trusts by any other name. Moreover, they were seemingly immune from the purview of any one regulatory commission, since their subsidiaries were headquartered in a variety of states. The primary result was artificially high utility rates. What trustbusters had been unable to do, reformers, including Lilienthal (and soon Swidler), sought to accomplish through more aggressive regulation, and, if necessary, by public ownership.[1]

Involvement in utility regulation ended up having important consequences for the direction of Swidler's career. In the early 1930s, he represented a Chicago-based consumer group before the Illinois Commerce Commission. Swidler's uncharacteristically acerbic comments about the way the hearings were conducted prompted the commission to cite him for contempt. Defending him against this charge was a stalwart of Chicago politics, Harold Ickes. The two men became acquaintances, which proved fortunate for Swidler. Franklin Roosevelt appointed Ickes as Secretary of the Interior, one of the New Deal's most important officials.

Swidler's relationship with Ickes, along with his admiration for Roosevelt, prompted his 1933 move from Chicago to Washington, D.C. While his stint at the Interior Department was brief, it was formative; hammering out rate codes for the National Recovery Administration provided him with valuable trial-and-error experience as a policymaker. In July 1933, just a few weeks after Swidler began work at the Interior Department, Congress established the TVA. Lilienthal, one of the three original directors of the TVA, recommended

Swidler for its legal staff. Given Swidler's support of the TVA's mission, he did not take long to accept the offer. But while the TVA's enabling legislation was fairly clear—the TVA was to sell power at wholesale rates, with priority given to municipal customers—there were few if any precedents to follow. At the outset, Swidler and other policymakers would have to create order from legal and administrative chaos.[2]

CHAPTER 1

Early Years, 1907–1930

The general drift of my story has been made familiar in scores of novels and biographical writings—the story of the leap of poor immigrants to a respected place in American society. I was born in Chicago in 1907 to immigrant Jews. My father was from a small town in Lithuania under Russian rule, and my mother was from Vitebsk, the capital of a Russian province of the same name. I had two sisters. Abraham, my father, worked as a tailor when he arrived in this country as a child of twelve or thirteen. By the time of his marriage, he was a partner in a small soft-drink bottling firm, Swidler & Bernstein. Dora, my mother, had worked as a seamstress before marriage. The family was never in want, but lived on the margin of economic security.[1]

My parents, who met and married in Chicago, were born about 1878 and arrived in this country around 1890. They were not sure of the exact dates, but had clear recollections of the Chicago World Fair (1893). Each had numerous siblings, all in this country. Some had arrived earlier or younger, had the benefit of American schooling, and were in better circumstances. A brother and sister of my mother were already rich. The rich relatives lived in Miami when I was growing up. Almost all of the others, on both sides of the family, lived in Chicago.

In my first few years at McLaren Grade School, I exhibited a language facility beyond that of my classmates. I spoke English with greater clarity and accuracy, was a better speller, had a larger vocabulary, and read books of greater difficulty. Thus was my career determined. In the culture in which I grew up, language talents were the province of lawyers. When I was no more than eight years old, my family took for granted that this was my destiny, and I accepted the goal without question, since I was unaware of any option that would make better use of my abilities. Throughout my schooling, courses were chosen for compatibility with a lawyer's role. Looking back, I wish I had been able to reserve a career decision until I was in a better position to make a choice, but arranged careers, like arranged marriages, have their advantages.

Similar to other Jewish boys in the neighborhood, I attended a Hebrew school (*Heder*) for a year or two before my thirteenth birthday, primarily in preparation for my bar mitzvah. The teacher was a middle-aged rabbi, a recent immigrant who spoke almost no English and knew little about teaching. He explained nothing, taught by rote, and scolded a great deal. He was not overly fastidious, and his clothes gave off a heavy odor. The ill-lighted living room of his home, filled with ranks of second-hand school desks, served as the classroom. He was misplaced, inadequate, and unhappy; he made no effort to disguise his feelings. I felt I was getting little out of the *Heder*, and resented the time I lost from play or reading. Nevertheless, I did passably well at the bar mitzvah services, but soon after entered the agnostic phase of my religious life.

The three Swidler children proceeded in turn from McLaren to McKinley High School, a four-year public school, some three-quarters of a mile from our home. For a year, the three of us, separated in age by a total of little more than three years, attended together. I entered in 1920. Here I continued to shape an identity by wide and indiscriminate reading, by writing for the school's literary monthly, and serving as a member of the debate team. The responses of teachers and classmates gave me a greater awareness of my personality traits and of my abilities and weaknesses as a student. I began to develop a social and political philosophy, and demonstrated an aptitude for expression in speech and writing.

As a product of my milieu, I became a liberal and a humanist, a cover-to-cover reader of *The Nation* and *The New Republic*, a critic of government institutions that failed to come to grips with the problems of racism, ignorance, poverty, and unemployment. (Sexism was not recognized as an offense in those days; I believe the word had not yet been invented.) These basic attitudes have not changed in the seventy-five years, more or less, since they were formed, although with time I have become skeptical of easy or complete solutions to social problems, and better informed on the benefits, as well as the limitations, of market forces. I discovered early that I was a fiscal conservative, an attitude that has endured. If I have lost some of my zeal for social reform, it was not so much because my ideas have changed in a fundamental way, but because passions tend to cool with age and ease of living, because American society has made progress in eliminating the worst abuses of the past, and because safe paths to further reforms are hard to discover.

My basic character and viewpoints were formed by the time I left high school in 1924, and were only more deeply set in the following six years of college and law school, from which I graduated in 1930. In intellectual orientation, I am largely a product of the 1920s. This was a period whose failures and abuses [later] called forth the humanitarian and economic reforms of the Roosevelt administration. Lynchings and burnings of African Americans were still commonplace in some parts of the South, and not unknown elsewhere. Throughout the South, strict segregation was the law and the practice, while in the North it was merely practiced. Sacco and Vanzetti, political radicals, were convicted of murder and executed in Massachusetts in a proceeding so lacking in fairness that millions of Americans were shocked by what they believed to be a grave miscarriage of justice. Strikes were violent, and sometimes the strikers were murdered by the police in the name of law enforcement. Jews were ruled out of employment in many large companies and law firms, and were not admitted by most clubs. Advertisements for lodgings frequently used the code word "restricted," which meant that Jews were not admitted. (It was taken for granted that African Americans would not apply.) Caveat emptor was the rule of the marketplace.[2]

At a comparatively early age, I began to follow the newspaper reports of the decisions of the Supreme Court, dominated in the 1920s by such arch-conservatives as Justices Van Devanter, McReynolds, and Sutherland, who found justification in the Constitution for invalidating much of the social legislation enacted by the states, and for whom "due process" protected property, not people. Chief Justice Hughes and, later, Chief Justice Stone, were swing votes whose views I respected, but my heroes were Holmes, Brandeis, and Cardozo, whose dissents pointed the way to the vast changes in constitutional interpretation that came about in Franklin D. Roosevelt's second term.[3]

College Life

My first college year, beginning in the fall of 1924, was spent at Crane Junior College, a municipal liberal arts college within easy walking distance of my home. Although I doubt that many of the student body came from wealthy families, the students were ambitious, alert, and receptive; the faculty members were competent, if not distinguished. It was not a bad place to begin a college education, and most of the students proceeded from Crane to a major university after one or two years. Among my classmates were Arthur Goldberg, Philip Glick, and David Krooth.[4]

At the beginning of the next academic year, my family moved to Miami. My father's failing business had closed, which forced me to pick another school for my second year. Reasons of convenience and economy dictated the University of Illinois, and I spent the 1925–26 academic year in Urbana-Champaign. It was my first experience away from home, and in the beginning I was lonesome and bewildered. The student body was one of the largest in the country. The campus buildings were spread over such a large area that it was hard to move from one classroom to another within the allotted ten minutes. Social life was dominated by Greek letter societies, rigorously organized on Gentile-Jewish lines. The food at the boardinghouse where I lived in the early months was not what I was accustomed to, and it caused me much distress. After a few weeks, I was hospitalized with what was first diagnosed as acute appendicitis. It turned out to be acute indigestion. I adjusted and survived, though not without difficulty. I disagreed with the elitism of fraternity and sorority life, but the need for friends (and better food and amenities) drove me to accept a bid. If dormitory accommodations had been available, I doubt that I should have done so.

My interview with Dean [Thomas] Clark in 1925 and the circumstances that led up to it are worth recounting. The University of Illinois maintained a ROTC unit in which all male students were enrolled during the first year or two of their studies. Although I was dubious about preparations for war and shared the prevalent cynicism about the causes and results of World War I, I did not at first elevate my objections to the level of basic principle, but instead enrolled with the rest of my classmates. I soon realized, however, that I was a pacifist. I declined further training on grounds of conscience. This stand was a shocker to the authorities, and I was summoned to an interview with Clark.[5]

I was awed, but stood my ground in the face of his statement that the regulations made no provision for exemption on account of conscientious objection, and that I faced expulsion from the university. In the course of the interview, during which Clark avoided questioning my patriotism, he asked me what I would do "if a Hun threatened to rape your sister." I avoided saying that I would stand by complaisantly, and responded that the event seemed too remote to require military preparations at that time. In the end, despite the absence of a provision in the regulations for exempting conscientious objectors, I was excused from ROTC attendance.

The incident is instructive, and I have thought of it often. With the ascendancy of Hitler a few years later, my views on military preparation and the views of the young people of my generation changed completely. We recognized that Dean Clark's question, however poorly framed, raised an issue of national defense policy that no longer seemed fanciful.

During my year at Illinois, the family business venture in Miami (a restaurant) had not gone well, and, at the end of the year, I was called home to help. I did not return to school in the fall of 1926, but stayed in Miami to work in the restaurant. In a few months it closed, a monument to the folly of entering a business without experience or adequate preparation. The failure of the restaurant freed me to return to school, and the family scraped together the means to enable me to spend the spring semester at the University of Florida in Gainesville, where tuition was free for Florida residents and expenses were small.

In many ways, the semester in Gainesville was idyllic. The campus was lovely and uncrowded. It was situated in a grove of magnificent live oak trees, trellised with silvery Spanish moss. The buildings were old but dignified and serviceable. The student body consisted mostly of native Floridians, "good old boys," who took their academic responsibilities lightly. The school relied on the honor system, which was administered flexibly, and few students failed their examinations. I saw a number of students bring their examination papers to their rooms and copy the answers from textbooks. There was something of a scandal during the semester involving a student's abuse of the honor system, but, as I recall, he was a Yankee and fair game. Class work was easy, and, in all but the last few weeks of summer heat, the weather was benign. The library was adequate and little used. I enjoyed four wonderful months of reading for pleasure, mostly in the great Russian and English novels.

I regret to say that it did not occur to me to investigate the Jewish community in Gainesville. There may well have been a synagogue or temple or a community center, where I would have been welcome. This was a period when I was not only rebelling against Judaism, but also against Jewishness, or at least against excluding myself from the non-Jewish world. I had not yet learned, as I did later in life, how to honor and enjoy my Jewish heritage, religious and cultural, while embracing the full opportunities offered by American society.

With the help of a relative, I was assured the minimum means necessary to go to law school in the next year. Since I intended to practice in Chicago, I applied to the University of Chicago Law School, which then admitted students with three years of college credits and awarded a bachelor's degree after the first year of law, by treating the law courses as the major subject for purposes of the undergraduate degree. I had only two-and-one-half years of college credits, but I had a reasonably good academic record and I was accepted, subject to a requirement (which proved to be a blessing) that I make up the necessary college credits before receiving my bachelor's degree. I was also granted some student aid in the form of a job in the law library.

I found summer work to help pay my school expenses. Indeed, I worked during every summer, and also during the school year, for the most part on weekends, and sometimes in the evenings, from the time I was a high school sophomore or junior until I had my law degree, the Urbana-Gainesville period excepted. In the last few months of law school, I found employment as a clerk for a lawyer practicing solo not far from the university, but I never found the ideal spare time job. I did everything from wrapping packages at a Sears store to selling shoes, operating a printing press, filling orders for a mail order company, teaching elderly ladies to drive a car, selling time by telephone for a radio station, and gathering real estate leads by telephone canvass. I spent one day trying to sell encyclopedias in a neighborhood where no one who spoke English responded to my overtures. I concluded that this was an unpromising field. I turned down selling insurance when I was told in an orientation session that the key to success was to sell to friends and relatives. Most of mine had no money for insurance, and I could not imagine myself imposing on them for commercial purposes.

Entering law school was a sort of fulfillment in itself, especially a school of such high reputation, but self-congratulation yielded soon to the press of work. As in the other leading law schools, instruction was based on casebooks that contained significant or representative judicial opinions, and other legal materials, as a basis for lectures and class discussions. I prepared a summary of every case as it was assigned, a substantial labor in itself, and maintained comprehensive notes on the lectures and discussions. It took most of the first year for me to adjust to the workload, and the first few examinations were filled with tension, even terror. No honor system prevailed; rather, all examinations were given under the alert gaze of assigned proctors.

When I entered law school in 1927, the faculty members were greatly distinguished, but elderly and, for the most part, not deeply involved in the legal currents of the times. Perhaps no other school could boast of so many celebrated names: James Parker Hall, an outstanding authority on constitutional law; Edward Hinton, a noted expert on evidence and legal procedure; George Bogert, the author of the standard treatises on sales and trusts; Ernst Freund, who had virtually invented the area of administrative law; Floyd Mechem, an authority on the law of trusts and agency; Ernst Puttkammer, a criminal law expert; and Harry Bigelow, the outstanding authority on the law of real property. They had been recruited from the great law schools of the East, and were well past their prime. Frederic Woodward, a somewhat younger professor, soon left the law faculty to become provost of the university. With the addition of a number of excellent young faculty, the school changed character entirely within a few years after I left. The process of change had touched only lightly the class of 1930.[6]

One of the young recruits in my time was Arthur Kent, and I took a course in constitutional law from him. The dynamism of the course and the effort to relate constitutional controversies to the changing problems and patterns of life in the country were a revelation. It was my best course, the first course to make the law come alive, and this from a man without a great reputation.[7]

In 1929, after making up my course requirements, I received a bachelor's degree and, in 1930, a law degree. To my considerable surprise, my law school grade average proved to be third highest in a class of more than 100, and I won the customary honors that go with high grades.

CHAPTER 2

Law Practice in Chicago, 1930–1933

The economic picture was bleak when I graduated from law school. The 1929 economic tailspin had worsened, and there was no sign of improvement. Banks were closing, business bankruptcies were peaking, tenants could not pay their rents, landlords defaulted on mortgages and tax payments, and many lost their properties to speculators. The economy was in a free fall. For some law firms, this was a time of prosperity and expansion. Bankruptcies and receiverships were lucrative areas of practice, collapses and price changes required adjustments in business arrangements by negotiation or litigation, and there was much legal activity in cleaning up the economic debris. Although not well recognized at the time, this process of readjustment was essential to recovery; to those involved, it was painful and frequently disastrous.

I had few options as to employment. I was offered a job by the neighborhood practitioner for whom I had done part-time work, and at the same entrance salary then offered by the downtown law firms—$100 or $125 a month. I turned it down as lacking in interest and promise. A girlfriend suggested that if we married, I could practice with her father and, in time, share his practice; the possibility of bondage to a staid practice (as I visualized it) diminished my ardor rather than increased it, and I did not follow up on this lead. Few prestigious firms, other than those headed by Jews, were taking on young lawyers. I applied to one of the largest Gentile firms, out of curiosity as to my reception. I was accorded a friendly and unhurried interview with a senior partner, but that was not yet the time to break the barrier to Jewish entry. Actually, I had no interest in working for a large law firm—Jewish, Gentile, or pagan. I can say now without apology that there were too many glaring injustices in the political and economic system for me to be content to seek success by uncritical pursuit of money or social position. A personal success achieved by ignoring the problems of the larger society did not seem a satisfactory or acceptable career for me. At the same time, I had a basic confidence in the American governmental system and in its capacity to cure itself. As a lawyer, I could help. I was determined to practice law in a way that would contribute to social reform, and I was particularly interested in two fields of practice—labor and public utilities.

Labor law was an obvious choice for a social reformer, and it attracted many young lawyers, both to government agencies and to private practice. This was a time of few protections for unions or workers, and, in most contests, the powers of government were arrayed against them. Anyone with strong underdog sympathies would find satisfaction in attempting to help workers to better their lives. The attractions of the public utility area to a romantic young lawyer in 1930 will not be nearly as clear to present readers. The utilities of that day, by and large, while rapidly improving their technology, were a source of corruption in government. Since they were monopolies and needed public franchises in order to build and operate their systems, they lent themselves readily to a process of corruption at all levels of state and local government. The scandals in connection with street railway franchises, in Chicago and elsewhere, were notorious. Statewide utility regulation was in its infancy. The earliest commissions, having been created in 1907, had not attained their present effectiveness, although a few states (Wisconsin, New York, California) were already doing a good job.

During the 1920s, I developed an interest in the movement for public ownership of utilities. A Chicago organization called the Public Ownership League proselytized for public ownership of electric systems. Its arguments were both economic and ethical. On the economic side, it argued that public systems could finance construction more cheaply because of the exemption from federal income taxes on the interest received by holders of municipal bonds, and that local governments could operate at lower cost than private systems. On the ethical side, the argument was that only public ownership would remove the incentive toward corruption at the local and state governmental levels. I attended a number of meetings of the League. Shocked and repelled by the utility abuses, I had no difficulty in accepting most of the case for public ownership. Only on the question of relative efficiency did I have doubts.[1]

Meeting Lilienthal

A well-known lawyer in Chicago, Donald Richberg, enjoyed a practice that embraced both of my interests, labor and utilities. The press reported some Supreme Court victories he had won on the public side in a utility case, and in railroad labor and telephone company rate cases. I read the decisions with special admiration for Richberg because his victories ran counter to the conservative tide of the Court. I sought out Richberg, applied for a job in his office, and explained my interest in working for him. He said he preferred to work as a sole practitioner. He added that a former associate of his, David E. Lilienthal, had established a similar practice and was looking for a young lawyer to work with him. I made an early appointment with Lilienthal and got the job. I had taken the fork in the road that determined my career, and began a relationship that lasted until Lilienthal's death more than fifty years later.[2]

Lilienthal explained to me the nature of the work in his office. A large part of it was editorial. Lilienthal was under contract with the Commerce Clearing House (CCH), a well-established publisher of legal materials, to create a new loose-leaf service for lawyers, *Public Utilities and Carriers Service.* This was to consist of a volume for Supreme Court decisions and a separate volume for each state where the demand warranted one. The idea for the set of volumes seems to have originated with Lilienthal. The volumes analyzed and summarized pertinent court and regulatory commission decisions, and listed each point that was decided at the proper place under a comprehensive scheme of headings and numbers. Only a few volumes had been completed, those for the Supreme Court and for New York, Illinois, California, and Wisconsin.

Editorial work was supposed to occupy half of my time, and law practice the remainder; in the actual mix, almost all was editorial. The work was pleasant and unhurried. It consisted of digesting the newly issued cases and incorporating the text of new laws and regulations on utility matters for periodic supplements to the existing digest volumes. We were also to make progress on additional volumes—initially, the Missouri one. The existing volumes reflected a high standard of legal writing and analysis, superior to that in the other loose-leaf services I had known, and I attempted to maintain the same level of excellence. While the work lacked excitement, reading and analyzing the new legal materials issuing from the Supreme Court, other federal courts, and from the leading state courts and commissions provided a quick education in public utility law and general constitutional law. I could hardly have achieved this in any other way in the same amount of time. For a period, I believe I was the world's leading authority on the decisions of the Missouri Public Service Commission.

Lilienthal became a pivotal figure in my professional life. He was only eight years older than I, a young thirty-one in 1930. Tall and handsome, he was also reserved and dignified. He was friendly, but not sociable. The door to his office was always closed and, for so small a group, there was a marked distance in his relations with [Harry R.] Booth and me. I doubt that we ever had lunch together. Once he invited me to Sunday dinner with his family at his suburban home, and he proved a good host. It was a pleasure to meet his wife, Helen, a woman of charm and strikingly independent views, and their two toddlers, Nancy and David. But that was the limit of our socializing.

Through Lilienthal I met Edward Morehouse, who became a lifelong friend. He taught utility economics at Northwestern University, and I had the experience of lecturing to one of his classes. At his invitation, I also did some writing for the *Journal of Land and Public Utilities Economics,* published by the University of Wisconsin. Just as the problem of the 1970s and early 1980s was the adjustment of regulation to conditions of inflation, in the 1930s the key problem of utility regulation was adjustment to conditions of depression and deflation. This was the theme of one of my articles.[3]

Lilienthal spent much time out of the office and left the CCH work pretty much to Booth and me. Aside from an initial hour, I do not recall receiving any instructions from him, and he never corrected my work, although I presume he reviewed some of it. In retrospect, it is apparent that the publishing and editorial venture was faltering; perhaps Lilienthal was also bored with it and was casting around for other opportunities. Utility regulation was then in turmoil because rates were inflexible, while other economic indicators—prices, wages and employment—were plummeting. It was a good time for a man of intelligence and imagination to be a utility commissioner. The problem was resolved in 1931 when Lilienthal received and accepted an invitation from Governor Philip La Follette of Wisconsin to serve as a member of the Wisconsin Public Service Commission, the utility regulatory agency of the state. Lilienthal settled out his contract with CCH and departed for Madison.[4]

Lilienthal's departure for Madison created profound changes in the relationship Booth and I had with CCH. The contract between Lilienthal and CCH made no provision for Booth and me. We soon became employees of CCH. For a short time, we worked in the LaSalle Street offices, but CCH closed them and transferred the project to its own publishing quarters, not far distant. The pleasure was now gone from the assignments. We were two of many, at small desks in a vast open space: a law digest factory. We were pressed for productivity, and no longer could take time for careful legal analysis. By mutual agreement, employment was terminated sometime in late 1932.

My First Partnership

Booth and I formed a law partnership after Lilienthal left, and when the old offices closed, we rented smaller quarters in another building on LaSalle Street. We had little paying business. Some exciting work developed, however, that not only provided interest and exercise in legal practice, but also gave hope of other rewarding opportunities. The principal activity was representation of a group called the Utility Consumers and Investors League. It consisted of a group of businessmen who were trying to hold down their utility bills, and included some public-spirited citizens who were interested in achieving an honest and orderly process for regulating utilities in Illinois. Prominent in the public citizen category were Paul H. Douglas and Harold Ickes. Despite the pretentiousness of the name, the group was small and poorly funded; utility investors in that day certainly needed protection, but I doubt that many investors were members.[5]

The case I remember best occurred in 1931 and involved gas rates in Chicago. Up to that time, the gas used in Chicago, as in most of the country, was manufactured from coal, sometimes as a by-product of steelmaking. Much of the gas associated with oil production was flared, and many natural gas wells were capped for lack of a market. Natural gas could be bought for a few cents a thousand cubic feet, and it was almost twice as rich in energy as manufactured gas. The enterprising [Samuel] Insull decided to bring natural gas to Chicago by means of a long distance pipeline, a pioneer venture in gas transmission.[6]

One would suppose that bringing in cheap and rich natural gas would be a great boon to Chicago consumers at a time when the city and its citizens were in financial distress. That, however, was not the Insull plan. His natural gas would not be used to supplant the local artificial supply, but only to enrich it, by bringing the energy level up from 530 British thermal units (Btu) per cubic foot to an intermediate level of 800 Btu. To insure that consumers would receive no benefit from the change, the retail rates were to be overhauled when the natural gas supply became available. The gas would no longer be sold on a volume (cubic foot) basis, but rather on an energy, or caloric, basis; the charge for the enriched supply would be the identical rate per Btu as for the previous manufactured supply. It was cruel that in this desperate period the people of Chicago were to be denied any benefit from introducing cheap and energy-rich natural gas to Chicago.[7]

The Illinois Commerce Commission (ICC) set the case for hearing without fanfare and on short notice. The press gave space to the technological features of the proposal, presumably using company handouts, but said nothing of rates. There was no opposition and no one intervened, except for the Utility Consumers and Investors League. I wish I could say we changed the result, but Booth and I were unable to do the trick.[8]

At the hearing, Peoples Gas lawyers put on their experts, who made a smooth and soothing presentation. There was no testimony in opposition, no questions by the ICC or its staff, and no cross-examination, except by me, and I was new to the case. The League had no money for economic or engineering experts, and neither Booth nor I had trial experience or significant knowledge of the gas business. The result was foreordained. I made an impassioned argument that affected the ICC not at all, and the project was approved unanimously. The League appearance and my interrogation did attract newspaper attention and drew some questions from reporters. In the exuberance of my new prominence and strong feelings, I intimated to a reporter my doubts as to the probity of the ICC members and the propriety of their ties to Insull. This charge made good newspaper copy. For this breach of etiquette, the ICC cited me for contempt. I should not have made the charge of improper influence in connection with an ICC proceeding in which I was participating, especially as

I had no proof. On the other hand, the facts were notorious—and the legislature had not granted the ICC contempt power.[9]

At the hearing before the ICC, I was represented by Ickes, whose participation guaranteed further press attention. Ickes was not one to pull his punches, and he made an effective argument, both on the law and on the merits of the gas case. It was decided that I would say I was sorry to have made the public statement, which indeed I was, and that the citation would be dismissed. I had not rescued the people of Chicago from Insull, but I had made a point that registered. This was at a time when it was unsafe in Chicago to voice criticism of Insull. I had also made a friend of Ickes, who was destined soon for a place on the national stage.[10]

During this period I was also a consultant for the Wisconsin Public Service Commission, at Lilienthal's invitation. Apparently, the commission was short of staff, and I was asked by Lilienthal to prepare drafts of decisions in several cases. Lilienthal seemed to like my work, and the drafts were used. I was compensated by the commission, though not extravagantly.

The Election of 1932

The upcoming presidential election consumed my interest in the summer and fall of 1932. I was a Roosevelt partisan. How could I not favor a man whose views were so close to my own? He favored strict controls of the utilities and a greater role for public power.

More than these specifics, I admired his character and style. His advisers were people of outstanding reputation. Some were even selected for their abilities and not for the usual political reasons. He was surefooted and made no serious blunders. He had grace and charm, a jaunty cheerfulness, and an aristocratic aura. Above all, at a time of economic collapse and growing unemployment, when the Hoover administration in Washington was paralyzed by indecision, timidity, and a cryogenic attitude on the role of government, here was a man who believed that government could help people climb out of the trough, who had ideas on how to do it, and seemed to have the courage, decisiveness, and ability to put his ideas into effect.

Roosevelt won the nomination and beat Hoover by a landslide. I was ecstatic. Of course, I wanted to be a part of the Roosevelt administration, which embodied all my aspirations. Also, I needed a job, since Booth & Swidler was having a hard time paying office rent. I had no political experience or connections and no idea how to go about seeking positions in the new administration—or at least not until Roosevelt appointed Ickes as Secretary of the Interior. It was a surprise appointment since Ickes was a Republican, but in view of his leadership in the conservation movement, a logical one. It took only a few weeks for light to dawn —record speed for me—and I sent a letter to Ickes expressing my desire to work for him in the new administration. I also solicited the help of Paul Douglas, who had become a close friend, and he reminded Ickes of my interest. Not long after, I received an offer to join the solicitor's office in the Interior Department. I lost no time in accepting. My title was assistant solicitor, a title shared by some other young lawyers. I reported for work at the Interior Department in late May 1933, during the celebrated first hundred days of the Roosevelt administration, in which much of the basic New Deal legislation was written, introduced, and rushed through Congress to the president for his approval.[11]

CHAPTER 3

Working in Washington, D.C., 1933

I took a Pullman car to Washington, but slept little. This was my first trip through real mountains, and the beginning of a new stage in my life. I peered through the window of my berth to glimpse the landscape in the moonlight and to watch the lights in the little stations as we rattled by. On arrival in the morning, I moved into the Brookings Institution, then located on Jackson Street a few steps from the White House. Brookings had lodgings for its visiting fellows, and Paul Douglas had made arrangements for me to stay there while I got my bearings.[1]

I was not destined to stay long in Washington, but in the course of a few months I moved several times. After a few weeks at Brookings and at a hotel, I joined several other young lawyers in renting a house in the Chevy Chase area. I no longer remember which of my friends of that period were also my housemates, except that Abe Fortas was one of them. It was a comfortable and stimulating environment. This arrangement continued until August, when I married Janet Voss. We lived in an apartment in a converted dwelling on New Hampshire Avenue until moving to Tennessee in November [1933].[2]

I reported for work in Washington on May 30, 1933, at the old Interior building at 18th and G Streets. The air in Washington was electric with activity. Many New Deal measures were already in effect, either by legislation or executive order. Ickes was at the center of much of this activity. He was the administrator of the NRA code for the petroleum industry. Also under his wing was the Public Works Administration [PWA], which had at its disposal vast funds for loans (and grants) to public agencies for the construction of public works projects.[3]

The solicitor of the [Interior] Department was Nathan Margold, by reputation one of the most brilliant lawyers in Washington. His deputy was Charles Fahy, who became an honored judge of the Court of Appeals for the District of Columbia. Fahy, originally from the East but long established in New Mexico, was another of the imposing legal talents brought to Washington by the opportunity to be a part of the Roosevelt revival.[4]

From Margold and Fahy I received a wide variety of assignments. Some involved traditional legal research. Did the secretary have the authority to prevent the post office building in Washington from being torn down? Could he transfer the inmates at a poorly equipped Indian hospital in South Dakota to the Saint Elizabeth facility in Washington? Others related to administration of the department. I remember reviewing the procedures of the Bureau of Indian Affairs and being appalled by the proliferation of reviews of simple matters of little moment. A file involving approval of the expenses of an Indian funeral could be an inch thick, and the cover copy could carry a dozen sets of initials, accumulated over a period of months. Before it came to the solicitor's office for final review, the signature copy of the letter of approval or disapproval was dog-eared and gray. As I recall, I recommended a drastic shortening of the review process and more delegations of authority for officials at lower levels to sign or reject such documents. The same story applied to approvals of actions within the Bureau of Reclamation.

I was also asked to interview the holdover lawyers in the solicitor's office and the various bureaus and to appraise the quality of their work for the purpose of deciding on their retention. This was serious business because careers and livelihoods were at stake. To start, these were not dynamic and aggressive men. Most of them, I presume, had the benefit of Republican political sponsorship, although I did not ask. In the interviews, they were deferential and obviously apprehensive. Jobs were hard to get. Although I was younger than any of them and had far less experience, I represented the victors in the election; they had come up in the spoils system and knew the rules of the game. My recommendations could help or hurt their lives. This was not a responsibility I relished. To my relief and surprise, these men (almost all government lawyers were males in those days) proved to be an able and hardworking group, accepting of old procedures, but not uncritical of them. They were desperately anxious to conform to the wishes of the new administration and to keep their jobs. I thought they could do good work under new leadership, and so reported. I believe all were retained, probably to their surprise.

The assignments to appraise some of the department's procedures and to evaluate the performance of some of the lawyers were disillusioning. My qualifications were modest, and the assignments seemed casual. Perhaps the press of business combined with the transition of the department to a Democratic administration after a dozen years of Republican rule allowed no alternative, but I would have felt better if the department had looked at the problems of improving procedures and personnel evaluation in a more structured and professional way. Ickes had some great talents, though orderly administration was not one of them. Indeed, with the growth of his power he turned into something of a despot, given to easy wrath and instant decisions. He even set up an internal employee surveillance system under Louis R. Glavis, who asked me to serve as counsel for his unit at a higher salary. I politely declined.[5]

The drafting and negotiation of the petroleum code under the NRA was one of the major activities of Ickes and the lawyers. Ickes's diary for the summer of 1933 reveals his focus on oil industry problems, especially the NRA Code. Aside from Ickes himself, the principal people involved were Margold and two young Yale Law School graduates, Norman L. Meyers and J. Howard Marshall. They had collaborated on an article in the *Yale Law Journal* on oil production and pricing problems. In the course of researching and writing the article they had acquired considerable oil industry expertise. I was only peripherally involved, but gained a sense of the process of code negotiation in talks with Meyers and Marshall, with whom I shared a large room. The oil companies believed they needed price controls and production restraints in order to make the commitments to labor required by the administration. Here, also, as in my own modest assignments, the process rested on the views of a few people, without benefit of a structured investigative or decisional process—and with little public input. The severe economic emergency perhaps permitted no more, though it was fortunate that the Supreme Court, in the *Schechter* case, rescued the administration from such an arbitrary and prescriptive way of determining the manner of carrying on the economic life of the country. This sage conclusion, I regret to say, did not become clear to me until years later. At the time I was too inexperienced and too caught up in the recovery effort to criticize the process.[6]

I had little direct contact with Ickes. I remember a few meetings on the top floor of the Interior building, in an office roughly the size and shape of a basketball court. Visitors were required to enter at one end of the room and take a long and intimidating walk to the Secretary's desk at the other side. I usually went as an assistant to Margold, who would present the problem, whatever it was; I doubt I had much occasion to do more than keep him company. Ickes always greeted me with a personal word; there was no occasion for more. I knew, however, that he had accepted my recommendations on procedures and people, and that he had relied on the legal memoranda that Margold had transmitted with his approval.

When my wife applied for work in the department, she was promptly employed. When I recommended two of my Chicago friends, Philip Glick and David Krooth (they had worked for leading Chicago law firms, but like young and talented lawyers throughout the country, were anxious to play a role in one of the most important and exciting experiments in the nation's history), he promptly found places for them. He was a distant figure, and despite these helpful actions, I had no reason to think he had any special interest in me as an individual.

The Tennessee Valley Authority

In the early summer of 1933, as a result of the creation on May 18 of the TVA, my life began to take a new direction. Roosevelt appointed Lilienthal to the initial three-member board of directors, and he took office in June. The other board members were two Morgans, Arthur E. Morgan, called "A.E." by the staff, and Harcourt A. Morgan—"H.A." or "Doc" to the staff. (Lilienthal was usually "Dave," even to people who had never met him.) A.E., as I shall sometimes refer to him to avoid confusion with H.A., was the first director to be appointed, and was designated chairman by Roosevelt.[7]

I was drawn into this pioneer undertaking when I received a telephone call from Lilienthal, asking me to join him in Knoxville as the "power attorney" of the TVA, a position that did not appear on the organization chart. In that capacity, he said, I would be responsible for the legal work in connection with the power program of the TVA, working directly with him in his capacity as a TVA director and general counsel, and serving administratively under the general solicitor, William A. Sutherland.[8]

The offer came as a complete surprise. Quite aside from any question of personal merit, the job involved legal responsibility for a large and critical new program. I was not only young but lacked experience as a practicing lawyer. While I was not long deterred by the risks, I took some time to determine whether my present position or the Lilienthal offer provided the greater opportunities. On the one hand, I was comfortable in my work at Interior and seemed to be well regarded. The Washington environment was stimulating. Living in the South seemed likely to entail difficult personal adjustments.

On the other hand, there was no theme or organizing principle for my work at Interior, and it was hard to be sure that my work made a difference. Many of the new Washington jobs involved loan or grant programs, and I feared becoming a bureaucrat whose professional life consisted of doling out favors from the federal government. I had long followed the legislative battles over the disposition of the Muscle Shoals properties, the heart of the electric power properties vested in the TVA. I was also aware of the importance of the multifaceted mission entrusted to the TVA, and of its broad powers to carry out that mission. Perhaps most important, I liked and admired Lilienthal. I decided that the invitation to work closely with him on some of the TVA's key problems was an opportunity not to be missed. After a few days, I accepted the TVA job, but made clear the need to delay the transfer until I completed my current assignments at Interior.

When Lilienthal told the other members of the board that he intended to appoint me, A.E. asked me to meet with him—I presume so that he could exercise his prerogative as chairman to advise and consent on appointments to key posts. We spent an hour together at the Cosmos Club, then located on Lafayette Park in a modest building at 15th and H Streets. Tall, thin, and dignified, Morgan looked the part of a college president, though he was stiff and formal. I think now that he may have been tempted to probe on the policy issues where he differed with Lilienthal, but thought better of it. His questions were directed to background and qualifications. Apparently I passed scrutiny, despite my past association with Lilienthal.[9]

Lilienthal exerted constantly increasing pressure on me to make an early transfer. He had negotiated a deal with the City of Tupelo, in northern Mississippi, not far from Wilson Dam, to sell power to its municipal system. The ceremony for signing the Tupelo contract was set for February 4, 1934, and the city was preparing a great celebration for the event. No work had been done on the contract. I was told that I must come at once to prepare this crucial document. Faced with these competing claims for my time, I made an early disclosure to Margold, who expressed surprise and regret, though not shock; we then negotiated a departure date. In the remaining weeks in Washington I made a start on Lilienthal's power contract assignment by collecting and comparing various models of the wholesale and retail power contracts used by private industry, as well as those employed by the Bureau of Reclamation and various other government agencies, including the Ontario Hydroelectric Commission. None of them fulfilled the TVA's special needs, especially with respect to controlling the resale rates of distributors, a purpose unique to the TVA, but they proved useful in framing the technical or formal contractual provisions, the "boilerplate" that sometimes receives little scrutiny. Since I was given no specific instructions or directions, the collection and study of prototype agreements was as far as I could go in Washington.[10]

A week or two before I was scheduled to leave, I asked to see Ickes so that I could say goodbye in person, to express my genuine thanks for his interest and to explain my decision to move to the TVA. Ickes declined to see me, and I was told by an aide that he would not be available. I was saddened and conscience-stricken. I had not realized that he expected more of me. I wrote him a goodbye letter, both apologetic and appreciative. In Knoxville I received a reply, castigating me as an ingrate. This eased my conscience. It was apparent that the problem was not one of protocol, but a demand for total loyalty—to the point of subordination of my own interests and career. On November 7, 1933, my wife and I, in a battered little Ford roadster, set out for Knoxville.[11]

PART II

Initial Work for the TVA, 1933–1939

Swidler and his wife adjusted quickly to their new environment in Knoxville. Being Jewish in the South could have caused the Swidlers problems, but it apparently did not. Though he himself never provided a possible explanation, the presence of a more visible and vulnerable group, African Americans, might have reduced the bigotry and discrimination directed toward Jews. Whatever the reasons, Swidler's recollections during this period focus on his professional, not personal life.

Numerous challenges confronted the TVA and its staff during the 1930s. Overshadowing all others was the conflict between the TVA and private utility companies. The latter either did not supply electricity, or supplied it at high rates, to Tennessee Valley residents. The TVA sought, within limits, to replace these private companies, and thereby increase the supply and decrease the cost of electricity. TVA officials believed that such a process could be amicable, primarily through equitable sale of private facilities to the government. The TVA would essentially scrap the generation facilities it acquired, replacing them with its own hydroelectric dynamos. It would use existing transmission lines where possible, while giving or selling the distribution facilities to municipalities. If private companies resisted this approach, then, as a last resort, the TVA could build its own transmission and distribution facilities.[1]

Executives and stockholders of the Tennessee Valley's private electric companies were horrified by the TVA's plans, often hurling charges of socialism and communism at the new agency. Wendell Willkie led these opposing forces. He initially seemed cooperative, even to the point of signing an agreement on January 4, 1934, whose terms called for transferring various parts of his Commonwealth & Southern utility empire to the TVA. This cooperation, however, proved to be a façade that quickly dissolved into antagonism. Willkie and other TVA foes supported litigation calling into question the constitutionality of the TVA's power sales.

If a desirable judicial outcome were not forthcoming, opponents still hoped that years of litigation would at least weaken the TVA's bargaining position if and when private utilities had to negotiate the sale of their facilities. Neither of the two most important cases provided these results. In *Ashwander v. TVA* (1936), the Supreme Court affirmed the TVA's right to sell any and all electric power produced by the Wilson Dam, reasoning that the government could sell its property (even if it took the form of electricity) in whatever manner it saw fit. In *Tennessee Electric Power Company [TEPCO] v. TVA* (1938), the District Court for Eastern Tennessee ruled that the government's commerce and war powers, as well as its right to dispose of property, made all aspects of the TVA constitutional. The Supreme Court affirmed that decision in early 1939.[2]

Though Swidler subsequently provided an analysis of this litigation, his part in preparing these cases was not substantial. Instead, he worked on equally important aspects of the TVA's development. First, he assisted in establishing an initial market for TVA power. He did so by helping set the TVA's power rates, writing municipal power contracts that became the model for almost all others, and by enforcing the terms set forth in the January 4 contract. Second, he helped draft legislation that released municipalities from the jurisdiction of state public service commissions, and authorized them to contract directly with the TVA (the majority in *TEPCO* cited this legislation to support its decision). Third, once a market foothold was secured, Swidler tirelessly traveled throughout the Tennessee Valley, persuading cities and counties to purchase TVA power. Indeed, the TVA's relatively low electric rates were contingent upon a significant expansion in the sale and consumption of electricity. This was not an easy objective to accomplish, since ongoing litigation resulted in intermittent injunctions, making if difficult for Swidler to know when and how to proceed. Litigation also undermined municipal willingness to rely on a government agency whose very existence remained in question until 1939. Finally, as Swidler notes, even publicly owned utility companies, desirous of maintaining local control, sometimes balked at the idea of buying strings-attached power from the TVA. The issue, then, was more complicated than the more publicized debate between private versus public power revealed. That Swidler was successful at all in this environment is a testimony to his skill and perseverance.[3]

Swidler negotiated contracts with industrial customers as well, another crucial factor in creating a market for TVA electricity. Perhaps the most important of these contracts was with the Aluminum Company of America (Alcoa). Swidler also played a significant role in the 1936 discussions between Alcoa and the TVA over the proposed construction of a new dam. Alcoa owned a large piece of property in Fontana, North Carolina. It was hoping to convince the TVA to accept this property in exchange for building a huge dam on it—from which Alcoa would receive cheap electricity for thirty years. Swidler opposed such an arrangement, and was asked by Lilienthal to present his objections to the TVA board. His presentation convinced the board to halt, at least temporarily, any efforts to obtain congressional appropriations for construction of a dam on the Fontana site. As Swidler rightly indicates, the board's decision illustrated a lack of confidence in the chairman, A. E. Morgan, who strongly supported the Alcoa proposal. Making matters worse, Swidler's presentation and the board's decision occurred at a meeting that was held while Morgan was out of town, a circumstance that deepened the growing rift between him and Lilienthal (Morgan thought Lilienthal had purposely arranged for the meeting to take place during his absence). No evidence, however, suggests that Lilienthal intentionally used Swidler to embarrass Morgan, or that Swidler even knew that the board meeting was going to be held while the chairman was absent. Nonetheless, Swidler's discussion provides a different perspective on the otherwise well-documented feud between Lilienthal and Morgan.

Still more TVA work lay ahead for Swidler. Between 1938 and 1939, the TVA began its most important negotiations with Willkie over purchasing TEPCO. For the better part of a year, no progress was made. Then, shortly after Lilienthal came down with undulant fever, Swidler and Julius Krug replaced him at the bargaining table. In early 1939, Swidler and Krug reached an agreement with Willkie, bringing a close to an almost six-year struggle between private and public power interests in the Tennessee Valley (the TEPCO purchase contract spanned over three hundred pages and included thirty-five parties). A *New York Times* headline echoed Lilienthal's assessment of the role Swidler and Krug had played: "Two Young Men Credited in Settling Power Issue." It would be one of the highlights of Swidler's career.[4]

CHAPTER 4

Bringing Power to the Tennessee Valley, 1933–1934

The trip [to the Tennessee Valley] was a fitting introduction to the problems of the region. As we moved west, we ran into rain, and in the last few hours, traveling through western Virginia and eastern Tennessee, we drove in a savage rainstorm. The crops had been gathered and the fields were clear or lightly covered with stubble. Much of the cultivated land was on hillsides. Thus exposed, torrents of red water were visibly carrying off the soil, in what looked like rushing streams of blood. I came from level, black-dirt country and was not familiar with the clay soils of the hilly South, or with the rapidity of erosion of farm hillsides in a hard rain. I was shocked to see the deep fissures and gullies, growing even as we passed. The sight was an object lesson in the desperate need for effective conservation measures before the soil was gone and nothing was left worth conserving. I resolved to try to make a difference.

When we finally arrived at our hotel in Knoxville, we found two messages waiting for us. We were invited to the Lilienthals' house for dinner, which was welcome, and I was told to prepare to leave the next day for Muscle Shoals to work with Llewellyn Evans, the TVA's chief electrical engineer, on the new wholesale power contract.[1]

In the 1938 congressional investigation of the TVA, neither Lee Evans nor I was asked to testify on the origins of the TVA rate schedules, although there was much interest in how they came to be. I was shown the schedules for the first time, in rough form, on this trip to Muscle Shoals. Lilienthal had expressed concern about Evans's rate structure, and wanted my views on their adequacy. So, I questioned Evans about them. He had not conducted a comprehensive assessment, but had based his recommendations on experience and judgment, supported by some quick studies by a young assistant, Edward Falck. Evans had sensible answers to my questions, usually illustrated with quick figures he jotted down on a pad. I remained concerned that the resale rates might prove too low, and proposed raising them slightly to provide a factor of safety, to which Evans did not object.[2]

Falck, whom I came to know well, and who was still a good friend more than fifty years later, provided important help to Evans. He studied the reports of the major public systems in the United States, including Tacoma, Seattle, and Jacksonville, to determine a relationship between costs and revenues. He made an estimate of TVA costs, which required at least a rough allocation of the common costs of Wilson Dam. He assigned half the costs to power and the other half to navigation and flood control. The special costs were assigned to their respective purposes (for example, the powerhouse to power and the locks to navigation). It was Falck who proposed the initial retail rate structure.[3]

It did not take long to develop a deep affection for Evans. He was then in his fifties, tall, spare, totally absorbed in what he was doing at the moment, but a man of many skills, wide interests, and naive curiosity about everything that entered his range of vision. He was a man of great sweetness. I never heard him say a critical word about anyone, and he accepted without sign of regret or concern the various administrative changes over the years that relieved him of managerial responsibilities and limited him to the role of adviser and consultant. Among other talents, he was an amateur sculptor, potter, and astronomer. He preferred vaudeville to a good movie—"They're alive." Simple in his tastes and unconcerned about status or appearances, he had the attributes of a Renaissance man, an inventor, and a thinker.

A great administrator Evans was not. Evans lacked the essential abilities of a good manager to set goals, analyze the workload, recruit staff, assign responsibilities, and supervise performance. In a conversation he always looked at the other party, even if he was driving a speeding automobile on a narrow and twisting two-lane highway, and even if his companion was in the back seat. I know, because I had a conversation with him in these circumstances. It was Falck who supplied technical support for Evans.

Evans' temporary offices were in the former Corps of Engineers barracks that had been built and used in connection with construction of the Muscle Shoals project. My principal interest was in seeing the Wilson Dam and powerhouse. These were of awesome dimensions. The powerhouse, however, was quiet, and the engineers in charge, with little to do, were subdued and gloomy. None of the generators was running, although a small amount of power was required for lights and miscellaneous purposes in the powerhouse, for the other facilities, and for the small town of Muscle Shoals, which adjoined the complex. This was a wet year, and the TVA was discharging enormous volumes of water through the floodgates, but none through the turbines. The Alabama Power Company had the right to buy power (at 2 mills per kilowatt-hour), but was not required to do so; it had enough power for its needs from its own plants. It also may be that the company had no desire to help the TVA look good by providing a market. The crowning irony was that with a million horsepower spilling over the dam, in a roaring torrent that seemed a mile wide and as deep as a tall house, the TVA's needs were so small that it did not pay to operate one of the giant generators. The TVA was therefore buying what it needed from Alabama Power Company.

The Tupelo Contract

I continued work on the [Tupelo] power contract when I returned to Knoxville. I conferred briefly with Lilienthal a couple of times, but he was preoccupied with many other matters, including the Willkie negotiations, about which more later, and the creation of a new agency, the Electric Home and Farm Authority (EHFA, or "Little Eva"). The contract was completed at the Shoals the night before the signing ceremony (and a great celebration, pictured in Lilienthal's book). It was delivered by messenger only hours before it was to be signed. Neither Lilienthal nor the Mayor of Tupelo had a chance to read it before attaching their signatures, but I suppose by then both had a pretty good idea what was in it. I did not attend the Tupelo ceremony, nor was I invited, and headed back to Knoxville.[4]

The Tupelo contract became the model for the contracts with 110 municipalities (including three counties) and 50 electric power cooperatives. Some of these agreements have already been renewed several times. It is a remarkable document in that it not only established the basic relationship between the parties on a long-term basis, but also incorporated—in a relatively few pages—a system of retail rate regulation for the 160 TVA distributors, which has withstood the tests of more than a half century. Nothing like it has been attempted by any other federal power marketing agency. I did not realize in 1933 how enduring and effective it would prove to be. Indeed, only after my years as a member of federal and state regulatory agencies, struggling to apply complex laws, regulations, and the accumulated case law of a century, did I fully appreciate the luxury and social economy of a system so spare and workable.

The wholesale rates of the TVA do not require elaborate explanation. They were the foundation for the retail rates, and it was necessary to dovetail the two sets. The most difficult issue was how to adjust for changes in TVA costs due to general changes in price levels. We decided that if the cost-of-living swung by 20 percent, up or down, either party could call for renegotiations. If the parties agreed on new rates, they would be put into effect; otherwise, a board of arbitration would set new rates. If the arbitration process faltered, rates could be determined by the aggrieved party, who could effect reductions or increases consistent with changes in the cost of living. Since many of the TVA's costs were fixed, it was unlikely they would fluctuate as much as the volatile cost-of-living index. I reasoned that the right of the TVA to increase rates in the same proportion as the cost of living would put pressure on the distributors to agree.

As a practical matter, the arbitration procedure has never been used, and there has never been a failure of agreement (although the negotiating process has sometimes been drawn out, and has always been controversial). Parties have acted on a common understanding that the results must protect the interests of both in achieving business-like results.

What I did not anticipate in 1933 or 1934 was that power costs could behave in a way that might be little related to general cost changes. This was because at that time it hardly occurred to me or anyone else, with the TVA's small market and an enormous hydroelectric construction program just beginning, that it would run out of hydro power and become dependent on coal and nuclear generation.

A Compact with the Utilities

References in Lilienthal's journals to the negotiations with Wendell Willkie and the terms of the contract that provided the TVA with entrée to major power markets are surprisingly spare. Lilienthal reports, in an entry dated June 23, 1933, that on June 16, at the first meeting of the TVA board, "a matter of particular interest" was a letter to Chairman Morgan from Willkie: "Willkie asked the chairman of the board to meet with him to discuss problems common to the Authority and his corporation. Then followed a considerable discussion as to what attitude the board should take toward the private utilities, etc. While fundamentally we may be in agreement, there was some difference of opinion as to tactics and strategy expressed as between myself and Chairman Morgan, with Harcourt Morgan acting as a mediator. This will require a good deal of working out."[5]

The next entry in Lilienthal's journal is dated almost six months later. On December 13, Lilienthal reports a conference with the president, who was sitting for a sculpture by Jo Davidson. After describing the scene, Lilienthal adds that "we got to talking about the plan that I had worked out with Willkie just that morning . . . [*ellipsis in original*]." Why Lilienthal limited himself to the atmospherics of his visit with the president, and omitted the substance, can only be conjectured. The entry is useful primarily in showing the president's interest in the TVA, and Lilienthal's care to clear important matters with him.[6]

Fortunately, Lilienthal included, as an appendix to the first volume of his *Journals*, an account of his conferences with Willkie, prepared in 1940 when the latter was the presidential candidate of the Republican Party. Lilienthal states that the account is based on "the files," supplemented by "sketchy" contemporaneous notes and, we may assume, his own excellent memory. In substance, Willkie first met with Chairman Morgan, and the two were in substantial agreement, presumably on the proposal Willkie made to Lilienthal after the board turned the negotiations over to him on August 9. Lilienthal and Willkie met in early October, and Willkie proposed simply that the Alabama Power Company [a subsidiary of Commonwealth & Southern] buy all the Wilson Dam power (at bargain basement rates) until Norris Dam was finished, which he thought would take six years. Lilienthal declined. They met again a number of times, working against a January 1, 1934, deadline, when the existing Alabama Power Company contract with the Corps of Engineers for the sale of Wilson Dam power expired. Lilienthal remained convinced that the TVA "must be given or would take" its own distribution market, and should either be accorded transmission rights to the networks of Willkie's companies or it would build its own.[7]

On December 12, less than three weeks from the deadline, Lilienthal and Willkie held two meetings, one before and one after Lilienthal attended an Executive Council meeting at the White House. At the first meeting, Willkie was adamant that he could not go beyond his offer to sell three distribution systems in northern Alabama. In the afternoon, Lilienthal, his back stiffened by a cheerful and supportive meeting at the While House, was able to persuade Willkie to go further. He was aided by learning from another official that Willkie was receiving bad news about a visit that Floyd Carlisle, chairman of the board of the Consolidated Edison Company of New York, had with the president. According to Lilienthal's informant, Carlisle had been "scared half to death" by Roosevelt's threats against the utilities. In the second meeting on December 12, and in a meeting the following day, Lilienthal and Willkie reached an agreement. The utilities would sell to the TVA all their transmission properties in certain counties in northern Alabama and Mississippi, their northern Mississippi distribution properties, and specific properties in Tennessee. After that, the negotiations "moved along very well with only relatively minor differences."[8]

Lilienthal's account refers to his Willkie meetings as stormy, and to Willkie as bullying and threatening. No doubt Willkie made a calculated effort to achieve dominance. Lilienthal described the negotiations to me at the time, in some detail, and I was fascinated by the difference between Willkie's private and public persona, a difference sophisticated people take for granted in politicians and key executives. At the meeting I attended, Willkie was a model of urbanity and friendliness in the earthy, country-boy style he affected, but the hard bargaining was over by then. There was a side to Willkie I did not know, and even when he sought to reveal a little to me of his broad views on social policy I did not take him seriously.

Willkie identified me correctly as someone of generally harmonious political views, and made several gestures in later years to engage my sympathetic interest, but I was too suspicious to respond, one of an unfortunate number of similar lost opportunities to learn more about the complexities and contradictions of the human character.

I came into the picture sometime in the last half of December 1933, when Lilienthal showed me a draft of the agreement. The major terms had been settled and the language of the agreement had been fleshed out, I believe, by Willkie and Lilienthal personally because it was too short to be the product of lawyers acting as such, rather than as executives. I made a few suggestions on the text, which Lilienthal (and Willkie) accepted, and attended the next meeting (in Knoxville), when the language was wrapped up. I wish I could claim greater credit because the agreement proved to be of great benefit to the TVA. I shall sometimes refer to this agreement by its date, January 4, 1934.[9]

The January 4 Agreement

The contract was for five years, until January 1, 1939, or until three months after completion of the Norris Dam power plant, whichever occurred first. Willkie was confident that Norris would not be completed for six years, but it actually took only half that time [1937]. Given a longer time frame, Willkie would have had a breathing spell, a period for maneuver and repositioning, before facing the threat of massive loss of market area.

Sales involving private utilities raised formidable questions of state/federal relations. The power companies were already regulated by state agencies. If the TVA attempted to regulate their rates for resale (of TVA power) while Willkie still retained ownership of distribution systems, a state/federal clash could hardly be avoided in any state that failed to withdraw its regulation in favor of federal jurisdiction. It became one of my early responsibilities to prepare drafts of authorizing legislation for the states. These statutes were written in broad terms, to cover any necessary financing needs, as well as authority to contract with the TVA. Companion bills covered the requisite authority for cooperatives. With the help of municipal supporters and sympathetic legislators, the bills were passed with little surface opposition. In the constitutional litigation, the existence of this legislation in the several states of the Tennessee Valley proved a crucial feature in the decision favoring the TVA.[10]

Willkie was looking forward to a Republican victory in 1936, and to a clipping of the TVA's wings before it could become entrenched, or before major damage was done to Commonwealth & Southern. The rapid completion of Norris Dam made it necessary for the TVA to start almost at once to build, through the heart of Willkie's Tennessee market, the long transmission link that would tie the two TVA dams [Norris and Wilson] together. This would greatly increase the TVA's firm power capability, make a power system out of two otherwise isolated dams, and provide a point of departure for transmission to the markets served by Willkie's Tennessee subsidiary, the Tennessee Electric Power Company [TEPCO]. Matters were not working out as Willkie had assumed.[11]

While the January 4 agreement was still being negotiated in November 1933, work started on a third large dam and power plant, the Joe Wheeler Dam on the Tennessee River [completed in 1937], near Decatur, Alabama. The Pickwick Landing Dam, on the Tennessee in Mississippi near the point where the river bends to the north, was started a year later [and completed in 1938], while Guntersville and Chickamauga [completed in 1940 and 1941, respectively] were begun some months after that. Plans were being made for the complete development of the Tennessee and its tributaries. It must have become apparent fairly soon that the Commonwealth & Southern strategy of sopping up the TVA's power output by turning over a small direct market area, and purchasing the surplus for the Willkie companies, could not work.

There was only a modicum of mutual trust to start with, and it soon dissipated. As early as March 29, 1934, I reported to Lilienthal that the negotiations between the Alabama Power Company and the Alabama cities [to arrange the latter's purchase of the former's distribution systems] were not going well. I said in the memo that "all conferences opened with the statement that the company must have cash, and would not consider bonds under any circumstances," at a time when it was known that the cities could not raise cash. Lilienthal protested to Willkie, threatening to help the cities obtain Public Works Administration money to duplicate the existing distribution systems. Doubtless, the threat would have been carried out if the companies had not yielded rather than see their investments destroyed, but there was risk that duplication would create a backlash. There would have been repugnance at the waste of resources and the losses to stockholders if duplication had become necessary. Even at the time it seemed odd, and seems more so in retrospect, that Ickes would lend himself to a pressure tactic based on the coercive effect of a program intended for a different purpose.[12]

Willkie and Lilienthal finally worked out an arrangement for the TVA to buy the distribution facilities at a lower price, then resell them to the municipalities. When the TVA made a tender to Chemical Bank and Trust Company of New York, however, the bank refused to release its lien (in the form of a mortgage trusteeship) on some of the distribution properties until additional financial terms could be met. The TVA regarded this action as a breach of faith, since the January 4 contract placed on the power companies the responsibility "to secure any necessary waiver of liens or encumbrances."[13]

The January 4 agreement held together until at least June 1, 1934, when the Mississippi Power Company conveyed to the TVA and local agencies various properties in northern Mississippi. A little later Willkie made clear that TEPCO would not sell the properties in the vicinity of Norris Dam, as provided in the contract. The TVA decided to make a formal tender [for these properties], and I traveled to New York with a check in my pocket on the Treasury of the United

States for $900,000, which I was afraid I might lose. That was a lot of money in those days, and I had not handled any four-figure checks, let alone one for almost a million dollars. I knew a thief or a finder could not cash it, but I was charged to make a tender the next day. Among other precautions, I used a safety pin to fasten the envelope containing the check to an inside coat pocket. But my tender was rejected, a clear contract breach.

Willkie had insisted that if the TVA was to acquire a market area of its own, the proper course was to exercise the right of condemnation, rather than to attempt to force sales by threats of duplicating transmission and/or distribution systems. Under condemnation, the price is determined in a court proceeding. Lilienthal regarded condemnation as a trap, and I believe rightly so, and rejected this course. The primary reasons were the uncertainties as to price, which would make financing difficult; the risk that the price in a jury proceeding would be largely excessive; and the certainty that such a proceeding, which would necessarily involve not only complex problems of valuation, but also statutory and constitutional questions that could not be resolved except in lengthy litigation, would cause delays the TVA could not tolerate.

To Lilienthal's reasons I added others. Since the TVA [originally] intended to operate only transmission facilities, it would be necessary to coordinate these purchases with the acquisitions of the various distribution properties by the respective municipalities. I saw no way to do this in a condemnation action; only a contract procedure provided the necessary flexibility. The long battle to bring the TVA to life demanded something more. Even with the cooperation of the companies, and with the benefit of specific congressional authorization, the difficulties were awesome.

CHAPTER 5

The TVA Becomes a Power System, 1934–1937

If the TVA had awaited the outcome of the constitutional litigation before adding new customers, the victories in *Ashwander* and *TEPCO* might have been barren. Expansion was essential to preserve the momentum of the program, to show gradually increasing amounts of power revenue, to retain the power staff by finding useful work for them, and to maintain the interest and support of the people in the area. It was also important to keep the benefits of TVA power and the success of TVA distributors before the public. Litigation itself focused attention on the TVA, and the nationally publicized controversies between Lilienthal and Willkie, and between Lilienthal and Chairman Arthur E. Morgan, served the same purpose. During this period Lilienthal proved a tireless campaigner. He had developed as a public figure, spoke often to groups in the Tennessee Valley, and was popular with local audiences in both urban and rural settings. He was proud of his success as a speaker, and considering that he was a federal official in the South—always the bastion of states' rights—and a Northerner at that, his achievement was remarkable. Without Lilienthal, there would be no TVA today.

Acquiring TVA Customers

Translating municipal interest in TVA power into willingness to sign power contracts was not automatic. A few cities were enthusiastic, but others were apathetic, or even antagonistic. Some could not be served for legal or practical reasons. To shape an effective marketing program, Lilienthal appointed a "New Contracts Committee," consisting of Edward Falck, then the head of the power staff, William I. Nichols, director of electric development, and me, as chairman. Nichols soon left to become editor of *This Week* magazine, and was not replaced. For the next year or two [ca. 1934–36], Falck and I jointly handled the new-business end of the power program.[1]

We thought that existing municipal systems would be so anxious to reduce their power costs that the eligible cities would line up immediately for contracts; they were, however, a resistant market. None had applied for TVA power, except Athens, Alabama. The support for the TVA came not from them but from the urban markets of the power companies and the unserved rural areas. They had much to gain and nothing to lose. The municipal plants, on the other hand, would surrender a degree of independence. They had fought off the acquisition efforts of the power companies over several decades, and, to some of them, TVA emissaries came as invaders. In addition to emotional considerations, they had a couple of substantive problems. The most important was the loss of control over surplus revenues. If the TVA had simply offered to sell power at bargain rates without attaching strings, the reception by the existing municipal systems would have been much better because some of these systems maintained high electric rates as a way of reducing or eliminating real property taxes. The TVA contract prevented such consumer exploitation, and the real estate owners of these towns, which included most of the influential citizens, had a stake in preserving the status quo. The mayors and aldermen also had questions whether the resale rates prescribed by the TVA would indeed pay the costs of system operation. The inertia of some of the municipal councils was a factor. They did not want to be confronted with the need for decisions, especially decisions to change fundamentally the way the towns had always been run.

Falck and I concentrated our initial efforts in Mississippi and spent many weeks on the road. In the early 1930s the Mississippi roads were not what they are today. Not all the main highways were paved, and those that were paved followed old cow paths, deer tracks, or Indian trails, full of sudden, abrupt, and poorly marked turns. The roads were narrow, and the ditches were close. There were no center or side stripes to help in keeping the car on the right side of the road or out of the culverts. The countryside was rolling and knobs hid the oncoming traffic. Automobiles were not as serviceable as those of today. Sealed headlights had not been invented, and night driving relied on two weak bulbs in fixtures easily penetrated by dust that confined the output to a glimmer. Night driving was an adventure. Both of us being young, we treated it as such. It should be added that the small-town hotels lacked the comforts of the modern chains. They were not air-conditioned, of course. Insect life was abundant, sometimes in the bed, and mice were ubiquitous. These discomforts were not unique to us, and we did not feel put upon. On the contrary, both of us enjoyed the experience as a whole and looked back on it with pleasure.

The effort to bring the small municipal systems within transmission reach of the TVA was productive. In Mississippi, Amory and New Albany signed contracts in the spring of 1934 and received service before the end of the year. Okolona signed a year later, and was served in July 1935. In Alabama, Athens became a customer in June 1934, and in Tennessee, Dayton and Pulaski joined the TVA system in the first months of 1935. In each case, both the signing of the contract and the initiation of service some months later were occasions for wide publicity. Loads were small, and even if taken together, had little impact on the TVA's operating losses, but no one could expect instant financial success, especially in the face of numerous lawsuits. They did demonstrate a continuing interest in the TVA program on the part of consumers that could be converted into a substantial market if the litigation were decided in the TVA's favor. At the same time the TVA was announcing these few additions, many other cities were signing contracts and making applications for PWA loans and grants. The effort to maintain momentum was succeeding.

During this period, TVA engineers considered the feasibility of building a transmission line from Wilson Dam to western Tennessee to serve a handful of independent municipal systems. These were Bolivar, Brownsville, Milan, Somerville, and Jackson. Economically, the line was borderline at best, since the distances were substantial and the loads were relatively small. Still, staff costs would continue whether or not the line was built, and employee morale in the power division would certainly improve with greater activity and the acquisition of new customers. Falck and I, assisted by a group of young TVA engineers—Walton Seymour, James A. Watson, and Edward Muir—set out to sign up these cities. The effort involved some tension because it was important to persuade all of them in order to make a decent economic case for building the line. For this reason, the contracts were not signed by the TVA as they were negotiated, but were held in abeyance until all negotiations had ended and the justification for the loop as a whole could be appraised.[2]

In general, the aldermen and municipal officials with whom we negotiated were difficult. Newspaper publicity had informed them of our intentions well in advance; perhaps they had consulted with each other and found a common base of skepticism over our answers to their questions. Could they depend on the TVA's longevity? The United States would not let them down. Suppose the TVA was found to be unconstitutional? The supply was from Wilson Dam, a national defense facility with high constitutional support. They paid nothing unless power was delivered, and their existing generating facilities could be reactivated if necessary. Would they be able to make money on the spread between the TVA's wholesale and retail rates? The engineers had tables, charts, and graphs for each to prove that they would, including the story of Tupelo's success. Moreover, the contract provided for an initial surcharge by way of precaution, and for rate increases if they proved necessary. Would the TVA dictate the details of operation of the distribution system? Only by setting the rate pattern and the flow of revenue, and by prescribing a uniform system of accounts. Each municipality could hire or fire as it pleased, purchase supplies and equipment without TVA oversight, do its banking, and determine its management structure however it wanted.

The heart of the argument made by the TVA group was the public benefits of the power program. Cheap power would save money for ratepayers, including cities themselves. Low rates and unlimited supplies would result in far greater usage, would brighten and revitalize the commercial areas of towns, and bring new industries and jobs. Cities would be relieved of the need to raise money for new generation units as the electric loads grew. The municipal systems would become larger, stronger, better managed, and greater contributors to the welfare of their communities.

At one night meeting, after the TVA team had made our case, without benefit of the usual questions that livened most proceedings, the city representatives took a vote and decided unanimously against contracting with the TVA. They said simply they were not prepared to make a decision. The TVA group could hardly believe what they heard. We were crushed and speechless for the moment. Recovering a little, I asked for permission to add a word. On the theme that the refusal to make a favorable decision was in itself a decision with fateful consequences for the city, I explained again the advantages of a TVA contract and why the risks were small and the potential benefits were large. The aldermen took another vote and approved the contract, again by unanimous vote. It was one of the high points of my career as a negotiator. That night the poker game, or perhaps it was craps, went on longer than usual.

In this as well as other meetings, the engineers presented the technical materials, while the legal and policy questions were left to me. Since the effectiveness of each TVA presentation depended upon creating and maintaining a favorable atmosphere and mounting interest, coordination was essential. My confidence grew with success, and, in turn, Lilienthal and the board developed confidence in me. Technical questions were left to the engineers, or talked out among us, so that there was never a minority report. It was a source of pride that the board never had occasion to reject a contract I recommended, or to require renegotiations of any feature. Gradually the people with whom I dealt—public officials, utility executives, and other businessmen—came to understand that I acted responsibly and that I had the trust of the board. I should add that over time the power division grew in executive strength, and that it took increasing responsibility for power contract negotiations.

As for western Tennessee, all but one of the cities [Brownsville] signed up. The board decided it would go ahead, considering that the investment could serve (as it ultimately did) as a basis for market expansion. All the contracts were signed on December 31, 1935. Soon, no municipal generating plants were left in the TVA area.[3]

Industrial Contracts

I also negotiated the early industrial power contracts. An industrial market was essential to provide revenues, improve the TVA load factor, and to demonstrate the favorable effect the TVA had on the regional economy. Willkie's fight against the TVA rested in part on the charge that there would be no market for the tremendous new supplies from the dams on the Tennessee River and its tributaries. The development of an industrial market was necessary to prove the validity of the TVA's assertion that the availability of low-cost power in the Tennessee Valley would create its own market. Soon after the *Ashwander* decision [1936], the TVA began receiving overtures from large industries. The first was Monsanto Chemical Company of St. Louis.[4]

Monsanto initiated negotiations with the TVA and TEPCO (Willkie personally handled the negotiations for the latter). Both were anxious sellers, and the Monsanto representatives played the two against each other with a practiced skill verging on artistry. We were told just enough to whet our interest and spur concessions, without learning the details of the TEPCO offers or the considerations that were animating Monsanto. The TVA's need for a market, and for a breakthrough in acceptance by large industries, was apparent. On Willkie's side, the considerations were equally compelling. He still had high hopes of victory in the *TEPCO* case, or from a turn of political fortunes. Furthermore, he could ill afford, in the heartland of his system, to allow the TVA a beachhead by serving the largest industrial load in the state, possibly excepting the Tennessee Eastman plant.

The TVA had developed a standard industrial contract, and we knew we could not make concessions to Monsanto and deny them to other customers. The concessions we did make were small compared with those demanded. The largest was to sell more than half of the total as secondary power, carrying a demand charge per kilowatt of thirty cents a month rather than the ninety cents for firm power. Secondary power was to be "available to customer for at least 300 days to be selected by Authority out of each calendar year in amounts up to 65 percent of the [total] demand." Minimum bills were to be calculated on the basis of 45,000 kilowatts of demand, but Monsanto had an option for another 5,000 kilowatts on the same terms. The contract was for ten years, with a renewal option for five or ten more years. In most other respects the contract was on standard terms.[5]

Willkie stretched further than the TVA and offered a contract at lower rates. What proved decisive to Monsanto was that his proposal could not be guaranteed for the full term. Under the state's police power there was always the possibility that the state, through its utility regulatory commission, would compel a change, especially for a contract that appeared to have discriminated against other TEPCO customers.

All in all, the TVA came out very well with respect to its bargaining position and the nature of its power system. It sold power on similar terms to several other companies in the next few years. The Monsanto contract held the line on the demand charge for firm capacity and on the energy component of its rates. There was good economic justification for a reduced demand charge for interruptible power. The prices yielded a reasonable return on the assumption of full utilization of capacity and thus complied with [amended] section 14 of the TVA Act, which stated that "in order, as soon as practicable, to make the power projects self-supporting and self-liquidating, the surplus power shall be sold at rates which, in the opinion of the board, when applied to the normal capacity of the TVA power system, will produce gross revenues in excess of the cost of production of said power." The breakthrough in the industrial market was a milestone for the TVA. It launched a period of rapid industrial expansion.[6]

The contract with the Aluminum Company of America (Alcoa) was not as simple. It represented a stage in a negotiating process that had antecedents going back almost to the turn of the century. Alcoa's engineers had initially identified the Little Tennessee River, one of the Tennessee River's main tributaries, as a potential source of large amounts of cheap hydropower. One location, the Fontana site in North Carolina, was wholly owned [by Alcoa] and held in reserve. By the 1930s, Alcoa had two goals. The first was to sell the Fontana site to the TVA; as payment, Alcoa would accept a large block of cheap—extra cheap—electricity to be delivered over a long period. The other was to reap the secondary benefits of such a plant. If the water releases from Alcoa's existing hydro plants and those of a Fontana site dam, whether built by Alcoa or the TVA, were coordinated with the management of the TVA power system, the combination would produce a great deal more firm power than the two uncoordinated systems. Alcoa wanted a large share of those benefits, including Fontana coordination benefits, preferably without having to make the large investment required to build a dam. Alcoa preferred to use its money to make aluminum and not electricity, especially if it could get the electricity at bargain rates.[7]

In the Muscle Shoals legislative battles, Alcoa had kept a low profile, but it left plain fingerprints on the TVA Act. Section 24 contained a special provision that authorized the president of the United States, within the three years following enactment, to purchase flowage rights or other property interests, and to pay for them by directing the board to deliver power from existing or future power plants "for a period not exceeding thirty years." This provision fit only

Alcoa. It was not slow in seeking to capitalize on its special status under the TVA Act. It had a choice whether to negotiate with Chairman Arthur E. Morgan, who as chief engineer was in charge of the TVA's construction program, or with Lilienthal, who was in charge of power sales. The transaction could be viewed as either the purchase of a dam site by the TVA or the sale of power. Whatever the reasoning, Alcoa chose to negotiate with Chairman Morgan.[8]

The Morgan-Lilienthal Rift

Chairman Morgan was wild to build the Fontana Dam. The dam would be the most spectacular in the TVA system. At 480 feet (as it turned out), it would be twice the height of Norris Dam, and taller than any dam east of the Mississippi. It would impound an enormous mountain reservoir. These impounded waters would flow through more dams on the Little Tennessee and the Tennessee River, through more feet of head, and do more work as the water fell to the sea, than the impoundments at any other dam. Morgan thought he was the right man to make the deal with Alcoa.

Lilienthal thought he had the better claim since the transaction involved the disposition of a large share of the TVA's output, and would vitally affect the economics of the TVA power program. He also believed, rightly in my judgment, that he could better protect the TVA's interests.

What was involved was more than a struggle for bureaucratic turf, although undoubtedly this was a factor. Basically, neither trusted the other. Morgan thought Lilienthal was a radical on the power issue. Lilienthal believed Morgan lacked good sense and was incapable of protecting the interests of the government or of the Valley's ratepayers in a negotiation with major industry executives. He also believed that Morgan was so anxious to build Fontana Dam—and to enlarge his role in the TVA power program at Lilienthal's expense—that he could not exercise whatever caution and restraint he might normally employ.[9]

Morgan argued later that he had picked a favorable time to negotiate with Alcoa, because Alcoa needed additional power supply, and its need for power provided leverage in negotiating for the Fontana site. Lilienthal asked me to appraise the handling of the negotiations, and I took a different view of the timeliness of negotiations. The reports I received indicated that Alcoa was insisting on a deal based on costs of 1 mill per kilowatt-hour for the power it received, and on a thirty-year contract. I thought this power value far too low and the thirty-year contract far too long. More important than these two features (on which I understood agreement had not yet been reached) was my belief that the objective circumstances made a favorable outcome for the TVA unlikely.

The bargaining took place in early 1936. The TVA was mired in litigation that threatened its existence. The litigation also prevented it from marketing its existing power supplies, and large new supplies from dams under construction would soon come on the market. The 1936 election created further uncertainty for the TVA's future. The board was divided, and unable to use to advantage its bargaining resources, such as they were. In these circumstances it would be foolish for Alcoa to fail to take advantage of the TVA's distress. I concluded that this was not the time to negotiate a transaction of such importance to the TVA's future. When the litigation was over, when the TVA had made greater progress in its power marketing program, and when the board could deal from strength, I believed a reasonable arrangement could be negotiated, but not until then. This view was confirmed by what little I could learn of the status of the negotiations Morgan was conducting. Morgan's anxiety for an agreement must have been as evident to Alcoa as to me, and apparently their position had not softened during the talks.

Had it occurred to me, I could have added another good reason for postponing negotiations with Alcoa. The TVA had neither the top-level technical staff nor the experience as a river management agency to negotiate advantageously with Alcoa about integrating the operation of the two power systems. Good engineers the TVA had in abundance, but except for the construction people, they had not yet been welded into a well-coordinated and efficient operating force. The TVA's power operating experience was limited to Wilson Dam, inherited from the Corps of Engineers. Norris, Wheeler, and other dams were well along in construction, though the practical problems of coordinated river management still lay ahead. Later, the negotiating team for the TVA was infinitely stronger and more experienced, and had a far greater wealth of staff support, than the best scrub team that could have been fielded by the TVA in 1936. Most of the individuals in the group that subsequently negotiated with Alcoa about Fontana were working for the TVA in 1936, in a variety of capacities, but they did not yet have the maturity and operational experience that were necessary to solve the formidable problems of joining two large river systems under separate ownerships into a single unit—one a private business and the other a federally owned corporation.

When I reported my views to Lilienthal, he concluded that the board should somehow arrest the drift towards making a poor bargain on such an important matter. He arranged for a meeting with H.A. on June 2, 1936, when, as it happened, Chairman Morgan was out of town. As I recall, in addition to the two board members, only the general manager, Jack Blandford, and I were in attendance. Lilienthal asked me to repeat for the benefit of the others the report I had made to him on the Alcoa-Fontana Dam situation. I did so, substantially as I have set out above, except with more particulars about the Morgan-Alcoa negotiations, the details of which have passed from mind. H.A. was clear that the chairman's determination to force a bargain at that time was imprudent, and that the negotiations should be discontinued. They voted to do so.[10]

This action was a stunning rebuke to the chairman, especially because it was taken in his absence. If there had ever been hope that the board could work together, the split on the Fontana negotiations closed the door. In the hearings of the Joint Congressional Investigating Committee there is conflicting testimony as to whether the meeting was scheduled during a brief or extended absence of the chairman. A.E. testified that the action came "while I was away for a day, down the valley at some of the dams." Lilienthal and H.A. had a different story. I have no recollection as to which was right—I assume all three believed what they said. But by this time the conflict between the chairman and the other directors had become a feud. The chairman had urged President Roosevelt not to reappoint Lilienthal when his term expired in May 1936, thus personalizing their differences in a way Lilienthal could not forgive. H.A. testified before the Joint Committee to this effect. In his testimony, A.E. denied this charge. He declined to elaborate or to answer other questions on this subject, and the committee refrained from ordering him to do so. I believe A.E. was standing on a technicality he did not identify, and that he did oppose Lilienthal's reappointment. The ability of the other board members to frustrate him was deeply embittering. At any rate, if Chairman Morgan had attended the June 2 meeting, the result would not have been different.[11]

As the hearings disclose, the chairman was a formidable record-keeper. At an early stage he must have decided to prepare himself for an eventual showdown, and he filled the files with his versions of meetings and other events as they occurred, especially policy debates. I did not attend the board meetings in this early period unless to make a report on my own work, and Lilienthal provided only elliptical accounts of what took place. It is evident from the hearings and other evidence that the meetings were marked by fundamental disagreement on power policies. The hearings revealed no dissension on the planning and construction programs that had been assigned to the chairman for direct responsibility; contention over the power program under Lilienthal's direction, however, was increasingly acrimonious.

Looking back, I can appreciate something of what Lilienthal endured for five years. When I was chairman of the Federal Power Commission, I was involved in a running dispute with another commissioner, Howard Morgan of Oregon (no relation to either of the TVA Morgans), which spilled into the newspapers and was ultimately resolved by a congressional investigation. To sit in meetings and know that anything you say may be used against you, sometimes out of context, in material that will find its way promptly into the press, is more than a source of acute tension. It is the sort of thing that interferes with sleep and the enjoyment of work and the other pleasures in life.

The chairman was accustomed to having his way, but Lilienthal and H.A. were formidable adversaries. H.A. was well known to the establishment figures in the Tennessee Valley, and enjoyed a solid reputation. Lilienthal had experience in public life, a quick mind, many well-placed friends, and a good record as a public servant. Because he was part of the board majority, he had the greater influence on appointments, other than for the engineering and construction departments, and came to represent the TVA mainstream. The managerial staff respected his judgment, while distrusting that of the chairman, and looked to him rather than A.E. for leadership. These were important assets, and Lilienthal used them to good effect. An equally important base of support was the group of congressional leaders who had fought hard and successfully for the creation of the TVA, especially Senator George Norris. They believed in public power and the yardstick, and they thought that A.E.'s support of Willkie's position was a betrayal of a basic purpose of the TVA Act. This group had excellent entrée to Roosevelt, who himself had a strong anti-utility bias. Above all, most of the people and press of the Tennessee Valley supported Lilienthal.[12]

A.E. was not without support of his own. The eastern press, including the *New York Times,* was in Willkie's camp and gave him prominent space. Willkie also seemed to have an outpost in *Time* magazine. The utilities of the country had influence with Congress and the press. The key, however, was that in a period of liberal political ascendancy, Lilienthal espoused the liberal cause. The contest seemed close at the time, but in retrospect it is apparent that A.E. could not win unless Lilienthal stumbled or Roosevelt was defeated before the TVA-Willkie battle was over, and neither occurred.[13]

Power Contracts with Alcoa

Even though the Fontana negotiations were postponed for several years, Alcoa still needed cheap sources of power in order to expand its operations. The first power contract with Alcoa [July 17, 1936] saved the TVA the cost of independent transmission to a new dam site in North Carolina [Hiwassee Dam], and the parties paved the way for possible later power sales. Either in connection with the first Alcoa contract or the next, I became acquainted with two men of importance in the TVA's later history. I was uncomfortable handling large industrial power contracts with my limited knowledge of power values and with the limited engineering support available in the power staff at that time. I began making inquiries as to the best available minds in A.E.'s division, particularly individuals familiar with power ratemaking. Several of the people I questioned brought up the names of two young engineers who stood out for their intelligence and capability, G. O. Wessenauer ("Wess") and R. A. Kampmeier ("Kamp"). When I talked with them I found that on their own initiative they had been making studies of the TVA's costs and rates, and had already reached some conclusions. It was a lucky day for the TVA. They inspired trust and their advice proved invaluable. In ensuing contract negotiations, I was guided by them on price, and felt comfortable with the results.[14]

Lilienthal badly needed these talents. After sounding out the two men on their willingness to change jobs, I suggested to Lilienthal that they be transferred from the construction to the power staff. He agreed and was able to arrange it. They became two of the half dozen most influential people in the development of the TVA power program, serving—with a third young man of equal stature, Walton Seymour—as assistant power managers. All were of about the same age. They worked as a team for many years, a truly formidable group. I never met their equal in utility management, on either the public or private side, except perhaps for Philip Sporn and his helpers in the American Electric Power Company.[15]

For the negotiations concerning the power contract of July 20, 1937, Alcoa sent from Pittsburgh its chief hydro expert and favorite engineer, the redoubtable James Growdon. Growdon knew what he wanted, which was a long-term electricity supply at less than two mills per kilowatt-hour. My minimum requirements were higher and I wanted a shorter term. It was not easy to get together, and we spent a couple of weeks in negotiation, but the transaction held a great attraction for both parties, and after each fruitless session, we agreed to meet again. I reported daily to Lilienthal, and consulted frequently with Wess, Kamp, and other staff as necessary. Growdon and I met alone, however. At some sessions little was said. He would sit silently, pondering and waiting for me to tire of the game and make concessions. I learned in negotiations to rein in my tongue and say nothing when I had nothing to say.[16]

A major problem for Alcoa was that it had waited too long. The TVA had sold large blocks of interruptible power, guaranteed for not less than three quarters of each year, to Monsanto and Victor Chemical Works. This quality of secondary power was already exhausted. The sale of more interruptible power so defined would have reduced the firm capacity of the TVA hydro system. Both Monsanto and Victor were assured of a twenty-year supply. So Growdon cried foul, stating that he could not buy on the same terms as Monsanto and Victor. Finally, with the background help of Wess and Kamp in getting over hurdles, we did put together a contract for the purchase and sale of 60,000 kilowatts, at a 100 percent load factor, and at the TVA's offered prices. Growdon finally accepted this formulation. Nevertheless, he won several concessions by way of compensation for TVA interruptions in the delivery of interruptible power.[17]

Once the ice was broken and a strong interconnection established, a slough of agreements with Alcoa followed. America was preparing for war, heavy industries were busy, aluminum was in short supply because of its importance for aircraft, and the company was under pressure from the defense agencies to increase production. The agreements provided additional power supplies to Alcoa and a degree of coordination between the two power systems. These arrangements were climaxed on August 14, 1941, a few months before Pearl Harbor, by an agreement under which the TVA and Alcoa hydro systems were married, but without change of ownership, except that the Fontana site was transferred to the TVA in partial exchange for a share of the benefits of integrating the two systems. I was assisted by Wess, Kamp, and Seymour in the negotiations. They were by now seasoned executives, fully capable of developing the substance of the agreement, and my role was appropriately that of the lawyer. We did not stand on ceremony, however, and they felt as free to raise or comment on legal questions as to interpose on the engineering and financial aspects. We had worked together closely on many matters, were good friends, and each contributed as he was able to our joint efforts. The standard joke among us was that they would quit practicing law without a license if I would give up designing dams and transmission lines.

The [Fontana] negotiations had an interesting side effect. As I traveled the winding mountain road along the Little Tennessee River [to Alcoa's Tapoco Lodge, where some of the Fontana discussions occurred], I noticed that Alcoa had made no provision for public access, unlike the TVA, which elevated public recreational use to a high status in overall river development. The TVA acquired and then set aside many thousands of acres along its reservoirs for swimming, fishing, and other river sports. The TVA donated these areas to state and local agencies for public parks, sometimes after the TVA had built cottages, beaches, and recreational facilities at its own cost. Not to wander too far from my point, it became apparent that a potential public recreational resource was being wasted by the failure of private licensees of hydro sites to provide for public use. Some twenty years later, when I became chairman of the Federal Power Commission, which issued the hydro licenses, I seized the opportunity to do something about it. Today all licensees must provide adequate facilities for public recreation. The change, no doubt, would have come about without me, but I hastened the day.

CHAPTER 6

Expansion of the TVA Power Market, 1938–1939

On August 25, 1933, only two months after its first board meeting, the TVA adopted a "power policy" as "the first step in carrying out the power program required by the [TVA] act." About two-thirds of the text was devoted to a statement of the TVA's intention to acquire an exclusive power market, its description, and the means of taking it over.[1]

In summary, the policy held that adverse effects on private utilities should not be a "determining factor"; that power should be sold at the lowest rates consistent with "sound financial policy" in order to accomplish the "social objectives" made possible by low rates; that initially the TVA program would be limited to: (a) the area adjacent to the transmission line the TVA proposed to build between Norris Dam and Wilson Dam; (b) the Muscle Shoals area (including northern Alabama and northeastern Mississippi); and (c) the region proximate to the Norris Dam (northeast Tennessee).

The policy also stressed that at a later time the TVA's service area would include roughly the drainage area of the Tennessee River in Kentucky, Alabama, Georgia, and North Carolina, and most of the rest of Tennessee. As a "fair measure of public ownership," the area was to include "several cities of substantial size." Finally, to avoid duplications, the TVA was to make a "genuine effort" to purchase facilities from private utilities on an equitable basis.[2]

The TVA adhered to this program with remarkable fidelity and success. That it achieved its goals in less than a decade—considering litigation, internal board dissension, and other obstacles—was equally remarkable. By the end of the major acquisition period, it was serving at wholesale an integrated area of 80,000 square miles, about twice the size of the Tennessee Valley itself. It did so through some 130 municipal and cooperative distributors in parts of seven states, including the major commercial centers of Knoxville, Chattanooga, Nashville, and Memphis in Tennessee; and Huntsville, Decatur, and the Tri-Cities area of northern Alabama. The only departures from the 1933 plan were extension of the TVA system into all of western Tennessee, and inclusion of a small part of Virginia, principally Bristol, together with an agricultural area in the upper Powell River valley.[3]

There was not a little exaggeration and showmanship in the public positions of Lilienthal and Willkie at this time. Despite Lilienthal's support and encouragement of municipalities to apply to the PWA for loans and grants with which to build competing systems, almost the last thing he wanted was to duplicate the utility properties and destroy the investment of stockholders. The thought of impairing the security of innocent stockholders and bondholders of the operating companies was deeply repugnant, especially to someone who intended to stay in public life, as Lilienthal did at this stage of his career. It was easy to visualize, if duplication actually occurred, the public revulsion over a period of years as the press printed pictures of redundant transmission lines and distribution systems. Without the threat of duplication, or an offer for the properties far in excess of their investment costs and the TVA's estimate of their value—Willkie would have been under no pressure to sell. Moreover, Lilienthal was not in complete control of the situation. The efforts of the cities took on lives of their own, and when a municipality voted overwhelmingly to issue bonds, and when it succeeded in litigation testing its right to do so, the letting of contracts and the erection of poles and wires could not be indefinitely delayed. Indeed, a minor amount of duplication actually occurred, enough to demonstrate that the threat was not empty.

All of the [acquisition] agreements followed a similar pattern. I shall discuss only one of them in detail, the TEPCO transaction, which was at the heart of the controversy with Willkie. In the January 4 agreement, the TVA's interests were limited to a modest amount of territory, a rural area in the vicinity of Norris Dam. As the TVA construction program developed to encompass the whole Tennessee River, including several dams and transmission lines in TEPCO territory, TEPCO became central to any logical operating area for the TVA. To be sure, no integrated service area was possible without acquiring TEPCO.

Negotiations for the TEPCO properties were initiated even before the Supreme Court ruled in the TVA's favor in *TEPCO* [1939]. In a public statement in early 1938, Willkie had offered to sell the entire system. Lilienthal responded with a written proposal that the TVA would buy not only TEPCO but also parts of the Alabama and Mississippi Power Companies, and that the price should be based on their actual cost less depreciation. Willkie replied that he was willing to sell these properties, though the price would have to be based on their "going concern" value, a term implying inclusion of many intangible factors and the absence of any realistic ceiling on price. Willkie also conditioned his willingness to sell on what he variously referred to as "stabilization" or a "territorial agreement," shorthand for provisions that would prevent the TVA from nibbling away at the properties Commonwealth & Southern retained.[4]

The question of a territorial agreement had been a stumbling block from the start. It was understandable that Willkie would not want to be in a position of fighting continuous rear-guard actions involving the properties of the Georgia, Alabama, and Mississippi Power Companies. On the other hand, Sections 10 and 11 of the TVA Act required the TVA to give preference in the sale of power to "states, counties, municipalities, and cooperative organizations of citizens and farmers" within transmission distance. Using hindsight, it appears that the conflicting interests—stability on the one hand, and municipal and cooperative preference on the other—could have been reconciled, with congressional assistance if necessary. But with enormous amounts of hydropower becoming available as the TVA construction program progressed, Lilienthal was unwilling to make any concessions that would preclude finding a market through sales to preferred agencies.[5]

After the initial salvos in early 1938, the parties agreed to assess the value of the TEPCO properties, putting to one side for the moment the other contentious issues. The TVA came up with a figure of $56 million; Willkie countered with $106 million. The gap seemed unbridgeable, but the parties persisted. Around this time, Lilienthal contracted undulant fever, which threatened to incapacitate him for months. He designated Cap Krug and me to carry on. Julius A. "Cap" Krug had come aboard about a year before. He was a year younger than I, and a brilliant, innovative, and forceful administrator who already had a remarkable record of achievement. With all this, he had a talent for friendship, appetite for adventure, and a youthful pleasure in camaraderie. Extremely tall and heavily built, he had a dynamic presence. We were pleased with the assignment, and to be candid, with the freedom that went with it in light of Lilienthal's desire not to be troubled during the painful stages of his malady.[6]

I believe the change in the TVA's negotiators was fortunate. The bargaining environment had been poor. In their dealings with each other, Willkie and Lilienthal were tense and distrustful. Willkie, perhaps partly to prove Lilienthal in the wrong, was gracious with Krug and me, and we did nothing to alter the mood. On the contrary, we tried to outdo him in politeness and affability, while bargaining hard on contract terms and numbers. The meetings were held in Willkie's office. His desk was always completely clean of papers, in contrast with his own rumpled appearance. A table at the other side of the large room was covered with books on national and world affairs. An article of his had recently been published by the *Saturday Evening Post.* He told me he had first offered it, without compensation, to *The New Republic,* but they rejected it. The *Post,* he said, had paid him $5,000, a large amount at the time, though it was evident that the money was not material. What he really wanted was exposure in a liberal journal. This was an early indication of the presidential virus; Willkie's open search for the Republican nomination did not come until a few months later, and I failed at the time to make the proper deduction from the evidence.[7]

On a disputed point, I rattled off some compromise language that Willkie apparently liked, but it was fairly complex. He asked me to get it down on paper. We proceeded to narrow our differences on money until we reached agreement on a figure of $78.6 million, subject, of course, to the usual adjustments for property added or retired after the base date, accounts receivable and payable, and unbilled revenue, which in the end added a few million dollars to the total. We then put to rest the details of the transaction, agreeing that the amount should be allocated among the TVA and the distributors, as they might agree, and that the TVA would assure that all, or substantially all, the properties would be conveyed. We were, of course, not writing the "definitive contract," but a memorandum of understanding.[8]

Willkie insisted that the TVA had adequate authority to finance the purchase by issuing bonds. He had the strong support of his New York lawyers, the firm of Weadock & Whiting, represented by one or both of the named principals, but usually by Weadock, who had been a judge in Michigan and was usually addressed by his honorific. Weadock looked the part of a judge turned Wall Street lawyer. Tall, solid, with a white mane and a handsome rugged face, he made an impressive appearance. He always wore vests with white piping and spats. Although age was beginning to tell, he was an excellent lawyer and Willkie deferred to him. I took strong exception to carrying out the transaction without obtaining new legislation. I was not persuaded by the arguments of Willkie and Weadock that the TVA Act, as it stood, was authorization enough. I had a lurking suspicion, probably baseless, that Willkie was laying a trap by proposing a course of action that invited litigation. The TVA was unlikely to win such litigation.[9]

When Lilienthal was well enough to talk about such questions, I discussed the matter with him, and he was in agreement. I was unable to persuade Willkie and Weadock, nor could they move me from my position. One of us, Lilienthal, Krug, or I, made the suggestion that we enlist the efforts of John Lord O'Brian, one of the country's greatest advocates and a revered figure among lawyers.[10]

A meeting was arranged with Willkie and Weadock, held, as I recall, in Weadock's office and attended by O'Brian, Krug, and me. O'Brian was at his best. He was less than five-and-a-half feet tall, normally soft-spoken, gentle and retiring in manner, but he cowed Willkie and Weadock before he was through. He used to dramatic effect the same arguments I had made several times to no purpose. Although he held no official position with the TVA, he said boldly that he would never permit the TVA board to proceed without new legislation. The meeting had not gone an hour before Willkie and Weadock agreed that the TEPCO acquisition arrangement should be subject to congressional approval.[11]

Now we were ready to write the actual contract. The scene was different from that in which the memorandum of understanding had been agreed upon by a few men in a quiet office. Instead, a large group took over the Commonwealth & Southern boardroom [in New York City], with two or three lawyers for each side, several executives, and a few engineers. Title and conveyancing questions were dealt with by a separate group working in Tennessee. We welcomed any of the numerous parties on the purchasing side who wanted to attend these deliberations, but only Chattanooga maintained a constant vigil. A large group of Chattanoogans headed by J. Wilhoite ("Jay"), chairman of the Chattanooga Electric Power Board, and States Rights Finley, its superintendent, sat in a circle of chairs along the walls of the room chatting quietly and making frequent trips outside to issue bulletins for the Tennessee press, or to consult with their bond lawyers and bankers. The Chattanoogans had no part in writing or discussing the contract, but their presence was a demonstration of the broad public stake in our work. Ultimately, the contract was signed.[12]

Obtaining congressional legislation [permitting the TVA to issue bonds for the TEPCO purchase price] was a feat in itself, and required the personal intervention of President Roosevelt. Fortunately, by the time the legislation reached the Hill, Lilienthal's health had improved to the point where he could—and did—take active charge of the campaign to secure its passage, which occurred on July 15, 1939. Congress gave its approval, and on August 15, 1939, the properties were transferred and payment was made. So many lawyers, bankers, trustees, and executives were involved that a whole floor of the First National Bank building in New York was cleared to accommodate all of the parties and the multitude of closing transactions. The press was in force, and three newsreels recorded the event. Lilienthal recorded the scene in his journals. He generously said that Cap Krug and Joe Swidler "really did the major part of the job in the past year."[13]

PART III

Washington, D.C., and Wartime Service, 1940–1945

The TVA entered a phase of institutional consolidation during the 1940s. The more prosaic job of managing an organization replaced the task of creating one. Swidler found this responsibility less engaging than the duties he had assumed in the 1930s. It is not surprising that he thought about trying to orchestrate a nomination to the Federal Power Commission; as he notes, though, he lacked the political support in Washington—and from Lilienthal—to obtain such a nomination.

Having to postpone his desire to move up the career ladder did not prevent Swidler from moving laterally between the TVA and other government agencies. His experiences in the Department of Justice, the War Production Board, and the United States Navy illustrate a number of significant characteristics about the expanding American state. As had long been the case, the federal government possessed a degree of permeability for mid-level civil servants. These individuals could and did assume a variety of roles in different agencies. Sometimes they served as conduits of information between agencies (for example, Swidler related information to Lilienthal about the nascent Atomic Energy Commission). They also witnessed conflicts—such as the "vesting controversy" Swidler describes between the Justice and Treasury Departments—that provide us with important insights about the nature of bureaucratic power itself. And, while Swidler does not explicitly address this issue, one senses that his work made him feel like part of a public service vanguard. America's mobilization for World War II—the way the federal government tapped into every aspect of domestic life—admittedly helped create a generation who felt greater civic purpose than previous ones had. But even before the advent of World War II, Swidler and like-minded colleagues thought about career advancement primarily in terms of continuing their public, not private, employment.

Remaining in the public sector did have its share of disappointments. Swidler felt isolated and underutilized when he returned to Knoxville in 1943 following his work in Washington, D.C. Prior to this period of frustration, Swidler had sought a commission in the armed forces; he was drafted instead. His initial excitement over being in uniform quickly gave way to dissatisfaction with the treadmill of desk-bound military life. Swidler also suggests that he encountered a degree of anti-Semitism among his fellow recruits, a refutation of contemporary propaganda that stressed the existence of racial and ethnic harmony on the home front. Swidler's transfer to the navy's Contract Termination Office provided a change of scenery, but did little to reduce his ennui. The military's necessarily autocratic orientation and its lack of intellectual stimulation confirmed, for Swidler, a preference for civilian public service.

Whatever his misgivings, Swidler's experiences during World War II put him in daily contact with individuals from very different walks of life. This was not only true while he was a Seabee, but also during his tenure in the Contract Termination Office. Most of Swidler's colleagues in that office were not enthusiastic supporters of Franklin Roosevelt, or, for that matter, the TVA. It is interesting to speculate what effect, if any, Swidler—who called himself an "authentic New Dealer"—had on this group of bankers and entrepreneurs. Perhaps none. Even if he did not make any political converts, Swidler played the role of loyal opposition, which likely had an educative function. The impression his co-workers made on him was more distinct. Their willingness to dole out public funds to various businesses troubled Swidler. More specifically, Swidler's memories of Lewis Strauss contrast with the latter's public image as a controversial and contentious figure, suggesting a more complex personality than commentators or historians have previously described.

What connected Swidler's experiences in the 1930s with those of the 1940s was his skill as a lawyer. This, in turn, emphasized how dependent the federal government had become on legal expertise. It was lawyers, more so than any other group, who helped define and redefine the parameters of administrative governance in mid-twentieth-century America.

CHAPTER 7

The Department of Justice and the War Production Board, 1940–1943

The victory in *TEPCO* and our acquisitions from Commonwealth & Southern in Tennessee, Alabama, and Mississippi were high points in TVA history. We were euphoric. The power program, the linchpin of the TVA enterprise, was on solid ground, and the opportunities for the future seemed unlimited.

Lilienthal decided that a celebration was in order and organized the closest thing to a junket in my TVA experience. Perhaps because it was so out of character both for the TVA and Lilienthal, it has stayed in my memory. The basic idea had some merit. Now that the TVA's principal responsibility in the power area was no longer to define its role and acquire a market, but rather to operate one of the country's largest power systems, the key people should visit other public power systems in the South and West to learn from them whatever might be useful. A small group made the trip, flying in one of the TVA's own planes. Krug, I, and the TVA chief engineer, Colonel Theodore Parker, were invited to accompany Lilienthal. The pilot of the plane, Tom Kesterson, was a graying World War I veteran and the owner of a flying service at the Knoxville airport. A genial companion and a careful pilot, Kesterson was one of the TVA family. The planes of those days were not as comfortable as later models. The cabins were not pressurized, the planes flew at relatively low altitudes, and travel was apt to be bumpy. The ventilation systems sometimes sucked in engine fumes. Some travelers were less affected than others. I was frequently nauseated. Safety systems were primitive, and the pilot frequently relied on visual guidance. On foggy days planes sometimes lost their way. Once, in a heavy fog, Kesterson flew into Memphis by following an electric power corridor from an elevation only a few feet above the transmission towers.[1]

The trip was both educational and enjoyable, but it had not been organized to take advantage of the opportunity to learn more about the power business.

Our first stop was in San Antonio, Texas. What I remember about San Antonio was a goat barbecue in our honor co-hosted by the [former] U.S. representative for the area, Maury Maverick [then mayor of San Antonio], and the hotel where we were put up. There must have been a couple of hundred guests at the barbecue, and it was a gay and friendly party. I learned that it required sharp teeth and strong jaws to bite and chew barbecued goat, and I never repeated the effort. The hotel boasted a presidential suite of Napoleonic proportions, which Krug and I shared. If we talked business, I do not remember it, but we enjoyed a pleasant stopover on our way to the West.[2]

We flew from San Antonio to Los Angeles, where Lilienthal left the group for local visits. I shall probably never forget the thrill of flying through the Tahatchapi Canyon. The wings of the plane seemed to brush the sides of the canyon. It was the kind of experience denied to modern travelers, flying 30,000 feet or more above the land. We did not see Lilienthal again until we picked him up in California on the way back to Knoxville. Parker, too, dropped out of the party. Krug and I proceeded alone, and had a grand time on visits to Boulder Dam, Bonneville, and the site of the great Grand Coulee Dam in eastern Washington.[3]

The trip took about ten days. When we returned to Knoxville, I faced a period of decompression and adjustment. The heady days of high-level strategizing and negotiation to build an empire against the opposition of the power companies were over. The era of improvisation, of instant response to challenges on the basis of visceral feelings and slide-rule calculations, was giving way to the era of the manager, who relied on a multi-level organization, careful preparation, and in-depth multifaceted studies. This would have been a good time to leave, and I thought about it. Apparently, Lilienthal had the same reaction.

During this period of transition, a vacancy opened up on the Federal Power Commission (FPC). I had good credentials—experience in utility regulation as well as an understanding from the inside of the dynamics of power system operations. My background was an asset in those days of public power ascendancy in Washington. What I did not have was support in Washington, not an easy matter to arrange from a Knoxville base. I thought Lilienthal could help me, and I talked with him about it. His first response was that when he was my age (he was eight years older), he had not yet held a presidential appointment. He asked why I should want to leave when the TVA was just coming into its own? I did not see the relevance of the milestones of his career, but it was true enough that there was much left for me to do at the TVA. It may be that neither of his responses reflected his true reason. Perhaps I had served so long as an assistant that he could not visualize me as an administrator in my own right. Some references to me in his *Journals* support this interpretation.[4]

Although I was disappointed, I had tried not to think of the FPC possibility as anything but a long shot. I was then the TVA's assistant general counsel, and I settled down to my work. I was thirty-two or thirty-three years old. When I was appointed to the FPC in 1961 from private law practice, I was fifty-four; I was fifty-nine when I left the FPC. In later years, I could not help wondering what my life would have been like if I had been able to begin my career as a senior Washington official twenty years earlier.

It was not long after the TEPCO closing that James Lawrence Fly, TVA general counsel, left to become chairman of the Federal Communications Commission. He was succeeded by William C. Fitts Jr., who had been serving as solicitor—in TVA parlance, the number-two man in the legal division. When Fitts moved up, I became solicitor. The two assistant general counsels were Thomas J. Griffin, an ex-Chicagoan, and Charles J. McCarthy, a native of Rhode Island. Fitts, who had practiced law in Birmingham [Alabama] in a small firm, was virtually a one-man law department. He could handle well, and quickly, almost any matter that came into the office, had less need to delegate than any other general counsel of my experience, and did more of the work in the office than could reasonably be expected of one man. Tom Griffin, who had much experience in corporate law in a Chicago firm, was a superb office lawyer, a meticulous craftsman in preparing complex legal documents, and an admired supervisor of the work of younger lawyers. McCarthy was a man of rounded legal talents, who thrived on litigation, and admirably complemented the work of Griffin.[5]

With these extraordinary legal talents at the TVA's disposal, and the TVA power program chugging along smoothly, I could no longer consider myself indispensable. I continued in direct charge of the legal work relating to the power division, but delegated a great deal to younger men. Under the circumstances, when I received attractive offers for temporary service of a kind that would broaden my experience and add interest to my professional life, I felt free to accept them if the TVA would grant me a leave of absence. Several such offers came my way, of which two were of some small historical interest.

The Department of Justice

The first was an invitation from Francis Shea, assistant attorney general in charge of the civil division of the Department of Justice, to work for the department as head of the Alien Property Bureau, which was one of the units of the civil division. The Alien Property Bureau at that time [1941] was still administering the properties seized from Germany and its allies in World War I. While the administered assets were substantial, the work was drawing to a close. A large, new bureaucratic prize had come into view—the responsibility for taking over (vesting) and administering German and Japanese assets if the United States should become a belligerent in World War II. These assets would, of course, greatly exceed the assets vested during World War I.[6]

The Alien Property Bureau had an obvious claim to the work, as an addition to the similar work it had performed and was still performing in connection with World War I. But the Department of Justice had a formidable rival, the Foreign Funds Control Office of the Treasury Department, which had recently been created to monitor and control the deposits and withdrawal of funds by the Axis powers. The Treasury claim to the new vesting responsibility had the vigorous support of the secretary of the treasury, Henry Morgenthau Jr., a good friend of President Roosevelt, to whom he had ready access. Frank Shea had heard of me, and sought my help in the ongoing jurisdictional battle. This seemed an interesting opportunity—resisting the Nazis, supporting Great Britain and the U.S.S.R. in their resistance, and preparing the United States for participation if need be. I accepted, contingent on working out an arrangement with the TVA for the loan of my services to Justice; I did so and moved to Washington.[7]

Paralleling the battlefield conflict in Europe, Justice and Treasury waged war for the vesting prize, which of course dominated our thinking perhaps more than the war in Europe, except when we were reading or listening to the news. During this period of uncertainty as to vesting jurisdiction, economic warfare problems were assigned to a three-man committee. It consisted of Dean Acheson, assistant secretary of state, who served as chairman; an assistant treasury secretary; and Shea, whom I served as deputy and stand-in. The meetings were held in Acheson's sunny and spacious office in what was then called the State, War, and Navy Building (now the Executive Office Building) adjoining the White House. I can no longer recall the specific matters that were discussed, but I retain a vivid impression of Acheson as a superbly gifted administrator. He was at ease and appeared unhurried, though the problems were crisply stated, the discussion pointed, and the meetings short and productive.[8]

The competition between Treasury and Justice ended in a draw. Perhaps to avoid the need to make a choice between two department heads for whom he had affection and respect, Roosevelt created a new agency [the Office of Alien Property Custodian in the Office for Emergency Management] to handle the vesting and administration of enemy property. He put Leo Crowley, a prominent Wisconsin businessman and political figure, in charge [in 1942]. Despite the draw, I felt I had not been a match for the Treasury team. They were better poised from their position in a vibrant new division than I was in a caretaker bureau. They started with a background in economic warfare, which I lacked. They were better oriented in the Washington official family. They may have out-pointed me also on brains and administrative ability. Still, I had hoped that the president would turn to the Justice agency, and we had made the best pitch that we could. I was greatly disappointed, but not surprised, when he failed to do so.[9]

The experience was not without value. I learned a little about economic warfare and a great deal about bureaucratic infighting, as practiced by Washington specialists in the art. Two lessons stand out. In the first place, I knew too little at the outset about the subject matter to make a running start, something of special importance in a job where the first weeks were crucial. Also, I had not adequately investigated the administrative complexities. Frank Shea hid nothing from me, but I had not asked enough questions. I certainly learned something about the difficulties of securing decisions at the top from a third or fourth level below in a large bureaucracy where I knew of no shortcuts. While I doubt that anything I could have done would have changed the result, I rate the Justice experience as a failure, on the whole.

Pearl Harbor was attacked during my tour of duty at Justice. Our preparations for war were even more provident than we had believed. I had missed the radio news of the attack and was reading quietly in my apartment on a warm Sunday afternoon when I was alerted by a telephone call from Robert Seaks (he was now executive assistant to Fly, who was serving at the Federal Communications Commission). I rushed at once to the Justice Department to see if I could be of any service drafting executive orders and/or legislation that would be needed in coming days. I found virtually all the lawyers from the counsel's office and all of the department's division heads already on the job, intending to spend the night. The atmosphere was one of controlled tension. I talked with Shea and determined what needed to be done in my own little area. I stayed on late into the evening when, running out of work, I reluctantly went home.

The War Production Board

In January or February of 1942, I returned to Knoxville, my leave having expired. I soon received a new call for temporary duty in Washington. I was asked by Cap Krug to take a job at the War Production Board (WPB). The problem was of the musical-chairs type. Krug had been moved up from chief of the WPB's Office of War Utilities (OWU) to a higher position. He had been succeeded at OWU by Edward Falck, but Falck had taken a special assignment. Herbert Marks, another of the TVA alumni, then serving as counsel for OWU, had been asked to fill Falck's shoes on a temporary basis. I was called upon to fill in for Marks, which I agreed to do if the TVA would again agree to loan me out. It was hard for any civilian agency to resist the defense agencies, so the TVA complied. By June of 1942, I was back in Washington.[10]

I remember especially the assistant general counsel of OWU, Fox Trowbridge, and a natural gas expert, William A. Lyons. A significant activity of this period was an assignment to realign energy use in an industrial area of the Midwest to conserve oil supplies so that more would be available for the military. With a small group, including Lyons, I set up a temporary office in Kansas City. We negotiated with the major suppliers and users, and were able to effect a substantial change in the energy mix by substituting natural gas and coal for oil. Under near wartime conditions, it did not require formal orders to evoke a spirit of cooperation. And I was impressed at how much easier it was to secure prompt decisions away from Washington and its many overlapping agencies and turf-conscious bureaucrats.[11]

When Falck and Marks returned to their old jobs, and my TVA leave expired, I returned to Knoxville. I missed the intense excitement of the work in Washington and the communion with people in the WPB and other agencies who were caught up in one of the most important efforts in our nation's history. By comparison, Knoxville was a backwater. Nonetheless, it had its own place in the national defense effort. For one thing, the TVA was building several dams at once, on record-breaking schedules, since power supply was the limiting factor in making aluminum, the principal material for airplane production. Prodigious amounts of electricity were needed for the Manhattan Project, a mysterious facility located in the hills near Knoxville. It was here that the nuclear materials for the atomic bomb were produced. The nitrate plant at Muscle Shoals was diverted from fertilizer to munitions production. In all, the TVA made an indispensable contribution to the preparations for defense and the possibility of war.[12]

CHAPTER 8

The Seabees and the Contract Termination Office, 1943–1945

After the United States entered the war, I sought a commission in the armed forces. I was almost thirty-five years old and had led a sedentary life. It seemed likely that I could be more useful as an intelligence officer, as a part of an occupation force, or in some other role that utilized my education and experience, than as an enlisted man in a combat group dominated by youngsters. I was encouraged by both army and navy officials, and took the necessary physical examinations, but the negotiations dragged on. Ultimately, I was told that I was disqualified by poor eyesight and by a new regulation prohibiting commissioning from civilian life. Soon after I returned to Knoxville from my WPB work, I was drafted. My mind was at ease, and I began to look forward to the adventure of service, preferably in the army. In June 1943, I was ordered to report to an induction facility in Chattanooga, where I found that my doubts as to my qualifications for combat were shared by the armed forces.

At the induction facility, there were throngs of men, most of whom seemed to be about half my age. After routine physicals, we were told to line up for assignment. The assignments were made by a non-commissioned officer, who devoted only a few seconds to each man. The men immediately ahead of me were divided between the army and navy, without any clear basis of classification that I could perceive. To my surprise and horror, however, I was assigned to neither combat branch, but to the Seabees. The Seabees! This was a bombshell. The Seabee division was the navy's construction branch. My experience was limited to the small chores of a householder. I protested and pleaded for an army assignment, but was forced to move along. There was no room for argument or any possibility of appeal. I became a Seabee.[1]

The Seabee division was unlike any other branch of the service in more ways than one. There was no construction training program. The navy relied on enlistees, and encouraged them by offering entrance to men with experience (there were no women in the Seabees) at elevated non-commissioned ranks, mostly seaman, third-class. The men were typically older than the combat recruits, averaging in their mid-thirties, and were cut from rougher stone than the draftees or the young combat volunteers. They were poorly educated and given to roughhousing and practical jokes. They were not a cross-section of any group except the younger half of white, southeast construction workers. They did not have a refined appreciation for northern Jewish professionals. I might have enjoyed their company had I been given an opportunity, which never came, to work with them in a small group and to know them as individuals. The Seabees performed heroically, especially in the Pacific theater, but I confess to problems of adjustment, although I managed to get along.

Boot camp for the Seabees in the southeast was at Camp Peary, in tidewater Virginia, on a flat expanse occupied by temporary military barracks and training grounds. I reported for duty a week or two after induction, bringing only the minimum of clothing and toilet articles, as instructed. At camp we formed long lines for issues of clothing, shoes, towels, and other necessities. It may be that Camp Peary was not as well run as other training camps. It certainly proved to be a haven for tyrants among the petty officers, who were virtually the only contacts for us recruits.[2]

Physical training was a mixed bag. I enjoyed marching and other drills, the small amount of pistol target practice we were allowed (my first use of firearms), and the crawling and vaulting exercises. I developed a charley horse in running exercises, though, and I lacked the strength in my upper body to scale a high wall by pulling myself up on a rope. The scaling difficulty was due in part to the fact that my left elbow suffered a multiple fracture when I was a child. I have never had full extension of it since. This defect was overlooked in the physical examination, and I did not mention it for fear of being rejected as a "4-F"—physically unfit for military service.

The biggest problem was with the nature of my work. I was classified as a yeoman, which in the navy means a stenographer or clerk. The camp had a sort of legal office consisting of a lieutenant and several yeomen, and I was assigned to this office. The primary function of the legal officer and his staff was to help the men on the base to protect their rights under the Soldiers' and Sailors' Civil Relief Act (SSCRA), a piece of wartime legislation that prevented divorce cases from being prosecuted while a man was on active duty. The officer in charge was also sought out for legal advice on other matters. He had no authority to represent the person involved. He could only refer matters to the federal district attorney or other officials, as appropriate.[3]

Fortunately, I had never cultivated any stenographic or clerical skills, having been more than content to leave such matters to others. The base had no facilities for training people in office work, or in anything else for that matter. The basic assumptions of the camp were that it served only people who came with the requisite skills, and that the purpose of the six-week stay at camp was only to familiarize recruits with military discipline and service conditions. At any rate, it soon became apparent to my superiors that my greater usefulness lay in counseling men on their legal problems, and I devoted much of my time to this work, which the legal officer seemed glad to turn over to me. The problems were reasonably diverse, and arose mostly under state law and the SSCRA. We had only one book in the office, a legal directory that devoted a page or two to a summary statement of the law in each state for a number of areas: the requirements for marriage and divorce, the statute of limitations for various offenses, definitions of various crimes, etc. It was not much of a library, but it provided guidance in a surprising share of the problems brought to me.

In addition to providing legal advice, I was assigned to give orientation lectures to new recruits. I gave a simple explanation of the provisions of SSCRA, so that they would be fully aware of its benefits. I was also told by my superiors to make clear that whatever else was true in civilian life, the navy was a dictatorship. I could never make myself transmit that message in quite those bald (and essentially false) terms, but found it possible to emphasize the requirements of military discipline nonetheless.

Several months passed in this not unpleasant way, and I watched the group with which I had entered, and several later groups, receive their orders and depart. Without explanation, I was not included. I then faced up to my own dim future in the Seabees. The work was becoming repetitive and boring. No sooner did I make a friend than he was called to foreign service. I decided I should try to get transferred from the Seabees to a navy activity for which I had greater aptitude and potential usefulness.

The Contract Termination Office

The transfer proved fairly easy. I had been in touch with Cap Krug. He had learned that the head of the Contract Termination Office of the Navy Department, Captain Lewis Strauss, was on the lookout for a lawyer. Krug spoke with Strauss, and Strauss requested my transfer (and commissioning, so that I would rank with the other men in his command); I received orders to report to the navy building in Washington. This was in the fall of 1943. Soon after arrival I received my commission as lieutenant, junior grade.[4]

The Contract Termination Office drafted the rules, set the standards, and monitored the performance of the procurement agencies. The group in Strauss's office turned out to be a dozen or so men, all but two of us Wall Street bankers or lawyers. The dominant person in the everyday work of the unit was John Young. For staff purposes, he ran the shop. I was evidently a puzzle to

Young and possibly to some of the others, since I was an authentic New Dealer in those days when Roosevelt was anathema on Wall Street. My conservative credentials were lacking and I was the dissenting voice on some policy matters. I did not attempt to hide my disagreements, a trait not always admired, even by me. In particular, I found it hard to accept the system for rating the performance of individual officials and offices in settling terminated contracts. The only formal criterion was speed. I thought consideration should also be given to the quality of the settlement in protecting the government's financial interests. Obviously, if a contracting officer is rated for speed, he will be tempted to buy speed by accepting contractors' claims at face value; furthermore, if the contracting company is aware that there is pressure to settle, it will decline any compromise of its claims.[5]

In this policy difference, the usual roles were reversed. The bankers, supposedly tight with their own and their clients' money, had little concern about spending government money to help business. As a bureaucrat, I was considered by the public to be profligate in initiating and carrying out government spending programs. It was not strange that I had doubts about issuing blank checks to contractors and subcontractors. The long fight with Willkie had primarily been about money. Had we been willing to pay the price, there might not have been need for controversy or litigation. Care in assuring that the government received its money's worth had become a TVA hallmark, and it was deeply ingrained in me.

Looking back, I am not sure I was right in my approach to settlements. A good case could be made that much depended on prompt settlements; if speed could be assured for a few billion extra dollars to assure that people could be put to work a few weeks earlier and civilian products made available a little earlier, the price was not excessive. But it was hard for me overnight to become a big spender of government money. And I was not satisfied that a reasonable effort was being made to introduce an element of economy that would not delay settlements. John Young, reflecting the big-business view of Roosevelt, argued that since everyone else was receiving a shower of gold from the treasury, businessmen should also. This argument was not reassuring. The important thing, however, was that the system worked, reconversion to civilian production was rapid, and the early postwar years were busy and prosperous. I cannot prove that the results would have been as good if I had persuaded my colleagues to accept my viewpoint of insisting on reasonable settlements. Nor could I have been sure that the navy procurement veterans would not have bargained for speed out of habit.

I had never had occasion to work with a group of people with such prominent names and prestigious associations. It was nice to be able to confirm that, like the rest of us, they had their good and bad points, that some of the richest and brightest were surprisingly unsophisticated about the world outside their ken, and that they exceeded the people with whom I had grown up primarily in having greater confidence in their own destiny. And why not? These were, after all, the most adventuresome and successful in their own world, and were most assured of family support to launch them in their chosen careers.

To be candid, in many ways life in the Navy Department was not much different than civilian employment, except I was in uniform, in a tight military hierarchy, earning far less money, subject to closer discipline, and with infinitely less discretion than in my other jobs. I lived in a cramped apartment and worked in a confined area of a large bureaucracy. In contrast with the TVA, the pace of work was unhurried, there was little occasion for travel, and we kept regular office hours.

I had good rapport with Strauss, saw him often, and he consulted me about many matters, mostly divorced from the subject of contract termination. He was a good friend of James Forrestal, undersecretary of the navy when I arrived, and secretary a few months later. Forrestal did not extend deference to his peers. He installed telephone taping equipment and, in the absence of any requirement to give notice to someone who was being taped, he secretly recorded many conversations. He was pleased and amused when a colleague in another navy branch said something gauche or uninformed. I observed his relish in playing some of the tapes.[6]

Although Strauss's role in the J. Robert Oppenheimer case revealed to the public a hard and uncompromising side to his character, his staff saw mostly a man of courtly manners, personal consideration, and private generosity. I had occasion to ask him a personal favor once. My mother, who had traveled to Miami for her health, found it impossible to secure reservations for the return trip to Chicago. During the war, trains were crowded, and Pullman reservations were hard to get. Strauss made a call or two and all was arranged, including my mother's limousine transportation to the railroad station.[7]

Strauss could also play the role of social host according to what he conceived to be the requirements of the event. He invited all the officers on his staff to a private dinner at his club on one occasion. After we had eaten, while we were enjoying cigars and liqueurs, he initiated the entertainment of the evening by telling a number of off-color stories, then encouraged others to follow suit. Whatever its appropriateness during a navy mess on the high seas, it seemed to me to be a strange way for a group of financiers and lawyers in Washington to spend an evening.

Unlike most people, who wait for opportunities rather than seek them out, Strauss was inner-directed, a common phrase at the time, and sought always to maximize his opportunities. Long before the war was over, he was considering what he wanted to do when it ended. He had apparently lost interest in making more money. When the war was drawing to a close, an opportunity arose that appealed to him. Washington decision makers saw the need to create an agency [the Atomic Energy Commission] to supervise and regulate the peacetime uses of nuclear energy. At this stage, Strauss, then an admiral, spoke to me about the possibility of Lilienthal's interest in chairing such an agency. He questioned me closely about Lilienthal's strengths as an administrator, and expressed his interest in serving as a member of the agency under Lilienthal's chairmanship. I told Lilienthal of these conversations when I returned to Knoxville. When Truman offered the chairmanship to Lilienthal, and referred in vague terms to a former banker who would be a commission member, Lilienthal was not at a loss for identification, but recorded in his *Journals* that this must be the man for whom I had worked.[8]

By the final months of 1945, the work of the navy's Contract Termination Office was nearly finished and I was anxious to return to Knoxville. In my absence, Fitts had resigned as general counsel of the TVA and I had been offered the post subject to my return, an offer I immediately accepted. Tom Griffin, TVA solicitor, was appointed acting general counsel. My appointment was by no means a foregone conclusion. Since the summer of 1941, I had taken two long leaves and had been in military service for two and a half years; needless to say, I was not up-to-date on the TVA's current legal and legislative problems. Griffin was a brilliant and diligent lawyer who had proved himself while I had been away. It would have been at least as logical to have given him the promotion. I do not know what considerations dictated the choice. I had greater longevity and higher visibility. Perhaps my military service was a factor. Perhaps the promotion was a reward for my work in building the TVA's power market. I was grateful to the board for the offer, which came as a surprise, and especially to Lilienthal, who must have had the decisive voice.

When I asked Strauss to release me, he delayed at first. Finally, he agreed to let me go if I would perform one more service for him. My assignment, he said, was to prepare an industrial mobilization plan against the contingency of another war. It did not seem to me at the time, or later, to be a sensible request, even assuming that Strauss was only seeking from me, and perhaps several others, ideas for a report he was preparing or considering. I argued that I had no substantial qualifications to prepare such a report; that no one person could know enough to prepare a worthwhile plan; and that I had no idea of the nature of the war for which preparation was being made. Strauss said to do my best, and that he would hold me for only two or three weeks. It would be wasted time, I believed, but this was the price of freedom.[9]

On a quick trip to Knoxville to buy a house, I found one that seemed suitable and bought it after telephone clearance with my wife. In December of 1945, my family and I returned to Knoxville, and I took up my new job at the TVA, which I was to hold for a dozen years, and leave in the midst of controversy.[10]

Postscript

In late 1942, my wife [Janet Voss] and I agreed to a final separation and divorce. The agreement was preceded by several earlier separations, for varying periods, and uneasy reconciliations. We had no children. After our final separation, Janet moved to Miami and applied for a divorce, which she received in mid-1943. Not long after, Gertrude Tyrna (called Gretchen) and I agreed that we would marry when we could live together after the war. My assignment in Washington enabled Gretchen to join me there, and in May 1944 we were married. We found an apartment on 15th Street, opposite Meridian Hill Park. In December, our daughter Ann was born.[11]

PART IV

General Counsel for the TVA, 1945–1957

Following World War II, America experienced unmatched economic prosperity, but also growing anxiety over a "cold war" with the Soviet Union. Swidler, as the TVA's general counsel, was not isolated from these trends. An expanding economy required the TVA to increase its power production, while concerns over communist infiltration prompted questions about Swidler's loyalty.

With its hydroelectric resources at peak capacity by the late 1940s, the TVA began to construct steam generator plants. These plants relied on coal, a fossil fuel whose exploitation scarred the landscape and polluted the air. Swidler himself seems to have been troubled by this development, though he never insisted on writing a legal opinion clarifying what, if any, environmental requirements the TVA could impose on its coal suppliers. In retrospect, Swidler recognized the historic constraints on environmental reform (pollution was seen as a regrettable, but necessary, cost of power production in the 1950s), without denying that the TVA's measures were "too little" and "too late."

The TVA's growing reliance on steam generator plants raised another issue. Like many government agencies, the TVA's funding came from annual (congressional) appropriations. Construction of steam plants, an expensive undertaking, required increased appropriations, but such requests incurred the anger of elected representatives who wished to limit (or eliminate altogether) the TVA's influence. Their views were supported by the more conservative economic milieu ushered in by President Eisenhower, who claimed that the TVA was an example of "creeping socialism." This line of thinking often translated into proposals to diminish the TVA's influence. Swidler played a role in opposing one of these attempts, the Dixon-Yates plan. His foot-dragging tactic during the initial discussions over Dixon-Yates provided TVA proponents with an opportunity to mobilize and articulate their opposition. Swidler's description of the episode illustrates not only his own views, but also an important feature of administrative governance: Officials from one political regime can exercise influence over the policy agendas of subsequent ones. Indeed, political scientist Aaron Waldavsky has summarized the TVA's opposition to Dixon-Yates as "a study in bureaucratic warfare."[1]

In 1951, Lilienthal, though no longer affiliated with the TVA, offered a proposal for making the TVA self-financing. Swidler, however, feared that Lilienthal's plan, which would have involved refinancing the TVA as a private (nonprofit) corporation, represented a conflict of interest. In short, the investment bank of Lazard Freres, for whom Lilienthal was then a paid consultant, would have likely profited by doing much of the TVA's refinancing under Lilienthal's plan. Swidler's concern on this point, undiminished after more than forty years, indicates lingering resentment (Lilienthal claimed Swidler did little to support his proposal, a charge the latter denied) and some regrets (Swidler believed that he should have been more forthcoming in his objections). Swidler himself supported a number of options for reducing the TVA's dependence on annual appropriations, eventually authoring the amendment giving the TVA authority to issue bonds.[2]

During these years, Swidler not only had to address issues involving the TVA's funding and autonomy, he also had to defend himself against charges of disloyalty. On August 8, 1951, the TVA's Loyalty Board accused Swidler of associating with communists and/or communist sympathizers; maintaining a friendship with James Lawrence Fly, reputedly a "concealed communist"; being a member of the National Lawyers Guild; supporting the Loyalist cause in Spain; endorsing the official positions of the Soviet Communist Party; and with arguing that the victims of the Soviet purges were martyrs to communism. Swidler could not, and did not, take these accusations lightly; he refuted all of them successfully. His first-hand account reveals the degree to which anti-communist crusaders and partisan politicians employed the rationale of domestic security to make reckless and unsubstantiated accusations. For the rest of his life, Swidler remained bitter over this incident.

In 1957, Swidler was forced to resign from the TVA. Ironically, his departure had little, if anything, to do with questions over his loyalty to America, but, instead, his allegiance to the new TVA chairman, Herbert Vogel. Swidler's own comments imply that Vogel's decision to dismiss him was not entirely capricious; nevertheless, Swidler's ability to retain his various positions in the TVA under four previous chairmen indicates more about his administrative and political skills than the fact that he locked horns with a fifth chairman. Furthermore, his reflections on the TVA during this period are not limited to self-analysis. He sketches a penetrating portrait of the TVA's little known, but highly influential Washington, D.C., liaison, Marguerite Owen. He also provides a fascinating glimpse into the personality of Lilienthal, who, in 1951, referred to his relationship with Swidler as his longest "professional friendship." Swidler does not conceal the mixed feelings he had toward his friend and former boss; in his opinion, Lilienthal's publicly articulated ambitions and, later, his seemingly uncritical embrace of big business, were incompatible with the values exemplified by the TVA and the Atomic Energy Commission. While at least one scholar has posited that Lilienthal's larger social vision remained consistent over the course of his career, Swidler himself would have found this proposition debatable.[3]

Swidler's speculation about why the TVA concept failed to generate support for similar domestic projects is also instructive. Swidler was one of the architects of the "Seven TVAs" bill, legislation that would have created several TVA-like "conservation" authorities throughout the country. Swidler's discussion of the factors that ultimately doomed this bill highlights an important element of New Deal state-building: In addition to ideological arguments, federal administrators—fearful that their agencies would lose power—undermined support for certain legislation.

CHAPTER 9

Service as TVA General Counsel, 1945–1957

Although I had been allowed much freedom in my work at the TVA, when I became general counsel, I had, for the first time, responsibility for a substantial group of people and for the legal affairs of an entire agency of diverse and important activities. It was a position that provided much pleasure and fulfillment, mixed with a little pain and frustration.

The legal division consisted of about twenty lawyers. All were located in Knoxville, except for a patent lawyer located at Muscle Shoals, Alabama, where the chemical engineering division was established. An important part of the work was the preparation of contracts and other legal documents. Virtually all legislation passed by Congress or the states affecting the TVA originated in the legal division. We also drafted amendments to exempt the TVA from general regulatory bills. On federal legislative matters, we coordinated closely with Marguerite Owen, the TVA's Washington representative, and it was she who was responsible for liaison with members of Congress. In meetings, she was frequently accompanied by one of the lawyers. Naturally, legislative matters were tied to policy questions and involved close coordination with the chairman, the board as a whole, or the general manager, depending on the circumstances. The heads of the respective bureaus and their aides were also active in these discussions. Many legislative matters were handled by Owen and the chairman, without further consultation, when the matters were free of legal complexity and policy differences.[1]

Marguerite Owen

The importance of Owen's role in the TVA has not been adequately recognized in other TVA histories, least of all in her own book. She had a key influence on policy in the Lilienthal and Clapp years. This influence rested on her understanding of the legislative process and her friendships with legislators and members of the White House staff, especially in the Roosevelt and Truman years. She was an effective legislative advocate, with a gifted pen for the kind of writing called for in her work, and a keen political observer. She was a woman of strong convictions and a stubborn will. Her place in TVA history warrants memorializing by more than a footnote.[2]

Owen was a native of Minneapolis and a member of a prominent family. Her brother was a leading member of the Minneapolis bar. Her roots and the prestige of her family were themes to which she frequently recurred with quiet pride. She never married, but enjoyed the company of men and gave many small dinner parties at her home. She did not invite single women to any of the several parties I attended. Although she never disparaged herself, I gained the impression she thought she was a disappointment to her family, despite her extraordinary gifts and accomplishments.

In a way, I was one of her favorites, which (I understand from others) did not spare me from caustic comments made in places where they could do me harm. Whenever I had occasion to be in the TVA Washington office, she invited me to visit with her in her private office; my presence provided her with an opportunity to probe for TVA Knoxville gossip, for which she had a sharp appetite. I told her what I knew that was not confidential, but this must have been frustrating because I did not gossip with anyone else, and to be a fruitful dispenser of gossip, one must collect from many sources. I might add that Owen herself was close-mouthed and volunteered little. I did not seek what she did not offer.

Owen had a tart tongue, but she guarded her talk and avoided saying much about public figures that could be quoted to her disadvantage. She was tough in her political judgments and in some of her political work. While she did not show it, I believe she was fundamentally a sensitive woman. She needed friends, which she cultivated, and I suppose she had as active a social life as most people in Washington. She never used vulgar expressions; she dressed modestly and in excellent taste; her parties, although informal, leaned towards punctiliousness of service and appointments; and she was a gracious hostess in an attractively decorated apartment.

Owen did her homework, knew her subject well, and was able to brief the TVA chairman before hearings and important meetings. She always played a background role, however. By her own choice, she never testified, and while she was a familiar of Washington columnists and newspaper people, and was generous in providing information to them, she did not speak for attribution, and usually preferred to direct inquiries to others on the staff or board. This attitude was of a piece with her general philosophy that within the organization credit should go to a staff unit rather than to an individual. She applied this view to the whole organization. Individuals in the TVA received no medals, plaques, awards, or other recognition—prizes were granted only to organizational units. This practice was either a valiant expression of egalitarian principle, or an extension to the TVA of practices adopted in her own life for her own reasons. At any rate, it obscured, and sometimes falsified, the realities of the workplace.

Owen herself had no assistant, or acknowledged none, and her personal secretaries—exceedingly able and devoted women whose work involved depth of knowledge, good judgment, and the exercise of initiative and discretion—were never entrusted with substantive responsibilities, and they were also kept at a social distance.

Before working for the TVA Owen had served in Washington as an assistant to Senator Bronson Cutting of New Mexico. She had many liberal friends in Congress: George Norris, Lister Hill, the La Follette brothers—Bob and Phil—and Hugo Black were among them. With many others, including most of the leaders in the delegations from the Tennessee Valley states, she established excellent working relations, though they seemed to lack personal warmth. Senator Kenneth McKellar of Tennessee, a powerful figure who sought unsuccessfully to dominate the TVA and who became a bitter enemy of Lilienthal, was in an arm's length relationship. John Rankin, a U.S. representative from Mississippi and an ardent public power advocate and white supremacist, needed no encouragement to support TVA measures. I am sure Owen attempted some discreet guidance in channeling his support, but she had little respect for him. Estes Kefauver and Albert Gore, as representatives and later as senators, were dedicated TVA partisans who worked well with Owen. She regarded both of them as lightweights, perhaps in comparison with Norris and Hill. In the 1930s, when Kefauver was still in the House, I testified on a TVA measure in a subcommittee hearing. When Kefauver, a subcommittee member, asked my opinion on a question of constitutional law, I prefaced my reply by referring to him as a "great constitutional lawyer who needed little advice on constitutional questions." While not entirely out of place in that setting, it was perhaps something of an overstatement. Marguerite thought it was ridiculous and never forgot or forgave this overblown tribute to Kefauver. She brought it up often to tease me, even after Kefauver's death.[3]

Lilienthal looked to Owen for ideas, not only those directly involving TVA management problems, but also for speeches and articles. In the preface to his book, *TVA: Democracy on the March,* he acknowledged a special debt to her "for the benefit of her rare critical judgment, for substantial editorial assistance, and many productive suggestions." Unfortunately, he provided no specifics, and her book is silent on the subject.[4]

Owen scored high in her ability to evaluate the Washington scene, but not all her ideas on TVA management were good ones. In particular, her campaign to have the TVA move its headquarters from Knoxville to Muscle Shoals was ill-timed, unsuccessful, and a public relations setback.[5]

Use of Revenues for Steam Plant Construction

Historians of the TVA, both the academics and those from TVA ranks, have given little notice to the battle in the 1950s over the TVA's authority to use its revenues for steam plant construction. The stakes were high. At the time, the existence of the TVA seemed to hang on the outcome.[6]

The question arose in the Eisenhower administration. The administration and the president himself thought of the TVA as a power enterprise whose growth should be rigorously limited and which should ultimately be transferred to the private sector. At any rate, appropriations for new steam capacity were out of the question. Furthermore, the TVA lacked bonding authority. Without being able to issue bonds, the TVA could not build for the future, nor could it be a dependable source of power.

With respect to using revenues from its power sales to construct steam plants, the TVA stood on favorable tactical ground. Congressional approval was not required and the TVA could determine for itself the extent of its authority. Under the Supreme Court decisions in the TVA cases, [private] utilities had no standing in the courts to challenge the TVA's interpretation, and the comptroller general could not disallow TVA expenditures. In a condemnation case, a landowner could contest the TVA's authority, but steam plants, unlike hydro projects, require only modest land areas. These properties could be purchased at negotiated prices.

This is not to say that the lawfulness of the TVA's steam plant program went unchallenged. On the contrary, the question was brought up in numerous congressional hearings, before both appropriations and legislative committees. Without the support of the TVA's friends in both houses, the fight would have been lost. Still, a major factor must have been the sheer unreasonableness of preventing a major power system, launched by Congress, from building a type of facility essential in almost every electric power network.

In these debates, I was the principal spokesman for the TVA. The board then consisted of General Herbert Vogel, an Eisenhower appointee, as chairman, and two Truman appointees, Harry Curtis and Raymond Paty. Vogel opposed the use of revenues for steam plant construction, but the others supported it. Neither Curtis nor Paty relished the responsibility for maintaining the TVA case in Congress. They were pleased to leave this chore to me. Indeed, the issues were primarily legal, and the general counsel was the proper person to discuss them. This created or intensified Vogel's animosity toward me. To my pleasure and surprise, when Vogel commanded a board majority and was in a position to determine TVA policy, he became proud of the organization and a strong supporter of its programs.[7]

Supreme Court Argument

Only one case was carried to the United States Supreme Court during my tenure. I presented the TVA's arguments. The case arose in North Carolina, and involved the TVA's ability to condemn a right-of-way in order to build a road that would replace another one flooded by a TVA reservoir. The issue was far from earthshaking, but it was important to the TVA at the time. I had never argued before the Supreme Court and was appropriately intimidated. I crammed for the occasion, and my argument passed without incident. The Justices were intimately familiar with the record, and able to summon every reported case bearing on the issues. Aided by the fact that the TVA's case was strong, and that the opposing counsel had overly fortified himself with spirituous beverages, the TVA won the case. What surprised me was the gentleness with which the Court dealt with my opponent, a country lawyer obviously beyond his depth.[8]

My Loyalty Proceeding

In the early 1950s I suffered through a loyalty investigation, an ordeal with many interesting facets. I had received security clearance up to "secret" for my work with the TVA, and "Q" clearance, the highest, from the Atomic Energy Commission (AEC), both after FBI investigations. It was a surprise, therefore, to be served with a loyalty complaint based on FBI information. It took me aback. There was no specification of anything I had done, good or bad, and no clue to the identity of any informant; there were only general charges that, on the basis of information believed to be reliable, I had consorted with communists and communist sympathizers, attended meetings of communist groups, etc. I was given only a short time to file an answer. I responded that there was no way to refute charges in such general terms. I therefore requested a specification of the facts that formed the basis for these allegations.[9]

Fortunately, this was the TVA. It had a high standard of fairness. Someone at the TVA bargained with the FBI for a specification of charges, and I was eventually served with a document I could get my teeth into. It did not name any informants, but it did provide some detailed allegations for me to address. Many people served with such charges resigned, either because of the difficulty of disproving them, or because of an association in the past that, publicized in the worst possible light, could have hurt their careers. It did not occur to me to take this course, because I knew that there was nothing even vaguely supportive of the allegations. The difficulty was in disproving the negative. Could I account for every moment in my life? Would anything else suffice?[10]

I should have had no fear, since nothing in my past tied me to communists or any other radical group. I had a middle-class and non-political upbringing. My great ambition was to be a lawyer and amount to something. I was not politically active in my university years, perhaps because I moved so much. I had never been a communist, had never been recruited or invited to join the party, and had only a nodding acquaintance with one person, a lawyer, known to me to be a communist. I did not participate in any parades, demonstrations, or other activities organized by communist groups. My career was unblemished, unless attacking the Insull utility empire was a fault, which I presume it was for the hunters of super-pure Americans.[11]

Nevertheless, I was afraid, which was true for millions of other Americans of my generation. How might an unfriendly stranger view my subscription to *The New Republic*? What of the books on my shelves? I tried not to be cowed and to be true to myself. This presented me with a stream of moral problems. After deep soul-searching, I canceled my subscription to *The New Republic,* on the basis that it had become predictable and boring, and that many issues were discarded unread after gathering dust for weeks. Could I ever be sure that I would have canceled except for a desire to give no basis for reports to the FBI by informants, possibly including the postman? Even an interest in Russia could prove troublesome. To my shame, I discarded several books, the titles of which I cannot recall, dealing with the Russian experience. True, I had more books than shelf space, but under other circumstances, would I have chosen these books for the discard box?[12]

The list of particulars on which my loyalty proceeding was based provided more information than the FBI intended, since it allowed me to identify every informant. None of them should have been given credence by the FBI. Furthermore, not one provided statements that, to a sensible person, should have seemed worthy of attention, even if believed. Some of them are included below:[13]

> EXAMPLE: Informant A stated that I had told him that the patent laws of the United States created monopolies and should be abolished. This tidbit could only have come from a patent lawyer for the TVA whom I had been required to terminate. The engineers and scientists at the TVA Muscle Shoals Research Center complained that he had refused to file for, or to prosecute, patent claims that, in other organizations, would have resulted in patents. When I spoke to him, he said the ideas were worthless. I looked at a few case files. In one he had actually filed a patent claim, but at a later stage conceded that the application was worthless. This seemed to be pretty bad on its face. I was not a patent expert, though, so I hired one to advise me, a former U.S. Commissioner of Patents. After an investigation, he reported to me that the only thing wrong with the TVA's patent administration was the lawyer. When

the lawyer was terminated and a new one employed, the TVA began to receive a stream of patents. Neither the original complaint nor the bill of particulars revealed any of this background. Instead, the former patent lawyer was referred to as an informant known to be reliable.[14]

EXAMPLE: Informant B said I had told him the end justified the means, and that for this reason the TVA had tried to entrench itself before the constitutional litigation came to a head. This informant could only have been the lawyer for the Knoxville power company that had sold its properties to the TVA, thus causing the loss of his best client and his subsequent bitterness. This conflict of interest was never mentioned. He had tried several times to engage me in a philosophical discussion of ends and means, and I had always changed the subject. He was referred to as a reliable informant.

EXAMPLE: Informant C was a lawyer for the TVA terminated by me in a reduction of force a year or two before. He had responded by accusing me of every possible fault from lack of diligence and poor management, to lack of legal capability—every fault, that is, except disloyalty. When the FBI questioned him on the loyalty question, he filled this gap with trivia, including an ex-secretary's work for the local Episcopal church. All of this, plus allegations of favoritism for, and friendship with, former TVA lawyers alleged to be associates of communists, led him to the conclusion that I must be a communist or communist sympathizer. The basis of this informant's animus was not disclosed. He was also referred to as an informant known to be reliable.

EXAMPLE: Informant D said he had overheard discussions in my home of the Soviet trials and the progress of the U.S.S.R. This man could only be the tenant of the apartment below mine in a two-apartment building. There was a good deal of discussion of Russia in the 1930s, especially at the trials of Stalin's former collaborators. Were their confessions true or false, coerced or voluntary, influenced by drugs, hypnosis, or promises of light sentences? The informant was hardly in a position to give a straight account of the discussions in my apartment, especially years after the event. None of this background was reflected in the report.[15]

EXAMPLE: Informant said she frequently attended meetings of the communist cell. In these meetings, Lilienthal was referred to as "Dave" and I was "Joe." I was reported to have said I was sympathetic [to communists]. This could only have been Mrs. Paris, who had virtually made a career of testimony before congressional committees on the communist conspiracy in Knoxville. In her most recent published testimony, she had said much the same thing, but added that I had never attended a meeting at which she was present. The latter bit of testimony, under oath, was not included in the particulars. Naturally, she was referred to as a reliable informant, and to some extent she was as to dates and cell membership. At least her list of attendees at cell meetings proved to be at least partially accurate.

EXAMPLE: An informant I could not identify said that I had attended a meeting organized by the communists to demonstrate support for the Spanish Loyalists. TVA travel documents showed I was not in Knoxville on the day of the alleged meeting.[16]

EXAMPLE: I was alleged to be, or to have been, a member of the National Lawyers Guild. It was further alleged that I gave a talk at one of their annual meetings. I acknowledged early membership and making the talk. I was able to prove resignation from the Guild when its political radicalism first became evident. I was also able to show (by an affidavit from Cap Krug, who accompanied me to the Guild's annual meeting) that Krug and I were occupied with TVA business that afternoon; our attendance was limited to the time required to make the talk.[17]

Taken at their worst, these allegations raised questions as to my ethical values and choice of companions, but they were far from casting doubt on my loyalty. Nonetheless, it was up to me to disprove the allegations as far as possible. This was a lengthy and arduous process. I was fortunate to have the help of two able TVA lawyers, Charlie McCarthy and Bob Marquis. I wound up, as I recall, with a hundred affidavits, some supported by extensive data. It was demeaning to be compelled to ask friends and acquaintances to provide statements about my good character and conduct. On the other hand, everyone I asked for testimony gave it gladly, always in full measure, excepting only Joe (Henry H.) Fowler, who did not answer my written request. He had served with me in Knoxville and at the WPB. His views on the genuineness of my efforts for the national interest might have had some value. Whether he doubted my loyalty, did not wish to get involved in a loyalty proceeding, or simply did not bother, I do not know. Affidavits aside, I admitted knowing three people who (it turned out later) were in fact members of the Knoxville communist cell. I described the extent of my familiarity, which was small, and said I knew nothing to suggest that any of them were communists or sympathizers. I also admitted that I had supported the Loyalists during the Spanish Civil War, and had been a financial contributor to their cause.[18]

The TVA Loyalty Board concluded that there was no basis for a finding of disloyalty. Its conclusion was confirmed by the Loyalty Review Board in Washington, which reviewed every loyalty file for possible error, whether or not there was an appeal. My security clearances were restored. A member of the TVA panel joked that I not only demonstrated that there was no disloyalty, I was also the only person in the TVA who had actually proved his loyalty. Naturally, I was pleased to end this painful episode. I knew, however, that a residue of suspicion was likely to remain among those who did not know me well. The fact that charges of disloyalty had been made by the FBI, and that a hearing was held, would forever be a part of my record. People would suspect that there must have been some substance to the charges or a proceeding would not have been brought. So far as other federal employment was concerned, I counted myself out. Why should an official risk criticism by picking someone whose loyalty had been questioned, when there were likely to be other qualified candidates about whom a loyalty question could not arise? I do not know—though it is a reasonable guess—that a part of my problems with Vogel when he became TVA chairman stemmed from the fact that I had been involved in a loyalty proceeding.

A few aspects of the loyalty episode seem to me to carry lessons, aside from the horrors of character assassination based on guilt by association. A shock to me was the mendacity of the FBI reports. The FBI agents in Knoxville could hardly have been unaware that their principal informants had scores to settle with me. In two cases, the animus of the conformants had exploded in the press. Yet the FBI said that all of them were "known to be reliable," with no mention of their reasons for bias. The agents lied. It was disturbing and frightening to be confronted, as I was initially, by vague charges from faceless informers. Had I not pressed for and received particulars, I do not see how I could have responded in a way that would have cleared my name.

The FBI agents and machinery seemed unsuited for this type of security work. Perhaps the FBI was capable of identifying and tracking spies, but they had poor qualifications for distinguishing between political liberals and radicals, or between unsophisticated youngsters (I do not include myself) and traitors or potential traitors. The agents I met were fundamentally white-collar policemen. They hid their ignorance in typical bureaucratic fashion, by covering their backsides and finding a basis for suspicion in every view and action not a part of the stereotype of a conservative WASP Republican. Better a thousand be falsely charged than that a single communist escape the net.

Before I was officially charged with disloyalty, I was interviewed by an FBI agent in the Knoxville FBI office. I was asked a few superficial questions—whether I was or had ever been a member of the Communist Party is the only one I can recall. There was not a single penetrating question, nor were there any questions that revealed the basis for the charges the FBI was about to make. The agent's attitude was formal and unfriendly. I sought to reach him across the chasm so that he could evaluate me as a person, but he would have none of it. Had he been seeking only to learn the truth, and had he provided me the opportunity to explain some of the incidents or statements on which the charges rested, I might well have dispelled his suspicions, or at least convinced him of the falsity of some of the charges and the unreliability of the principal informants. He might thus have avoided initiating a proceeding that gave me great pain and inflicted lasting damage, was costly to the government in terms of wasted effort and distraction from useful employment, and that did the FBI no credit. The only purpose of the interview was to develop a case of perjury should I eventually be caught in a lie. I thought I was entitled to more than an effort at entrapment from an agency of my own government.

Finally, I believe that the motivation behind the FBI's effort to entangle me, as well as the extravagant attention they paid to a pitiful group of politically naive young people, was part of an attempt to discredit liberals by labeling some of them as communists or sympathizers. What proof do I have of the existence of such a malign campaign? Not a great deal now, but at the time there was not a little speculation to this effect in the press, based on [J. Edgar] Hoover's known views and the actions of the FBI; Hoover's total support of McCarthy and McCarthyism, animated by a hatred of political liberals in general and the New Deal in particular; the reactionary Republican press's support of Hoover; the direction of the loyalty program; and conversations at the time, too vague in memory to attempt to detail, with people who were caught up in a dragnet one way or another. Unfortunately, I made no contemporary notes of these conversations. I do not seek to persuade the reader, but only to record my views.[19]

CHAPTER 10

Environmental Protection and a Privatization Effort, 1951

One of the early demands on the TVA was that its coal contracts should stipulate mining practices that limited harm to the landscape. The major offenders were the auger mines on the hillsides and the surface mines, familiarly called strip mines. The auger mines were small mines that generally followed an outcrop of coal around a hillside. The coal seams were typically only a couple of feet in thickness, sometimes less, and they were mined by driving augers into the seam to loosen the coal and bring it out. Production was pitifully small, but such a mine provided a living for one or a few men until the seam was exhausted or could no longer be reached by the auger. To reach the seam, it was necessary to build a road by grading along the slope. Even a primitive road involved moving large volumes of rock and dirt, usually by dumping down the hill. Drilling also involved excavating much waste material, which was disposed of in the same way. The loose dirt clogged the valleys and poisoned the streams, all for a meager recovery of coal.

It seems incredible now that the TVA would do nothing to discourage auger mining. I talked with Kampmeier about it at the time. He was the deputy manager of power and the presiding genius in coal procurement (although the paperwork was performed by the purchasing department). He was one of the most capable, ingenious, and quick-witted engineers I have known. He was also persistent and stubborn—an ideal man to have on your side in a negotiation, and formidable in opposition. He was so aggressive in holding down the price of coal to the TVA (and elsewhere by contagion) that he became the *bête noire* of the coal industry. I asked how he could justify encouraging auger mining in light of the long-term environmental devastation it caused, all for a paltry amount of coal. He said that if auguring the coal was the most profitable use of the land, it was economically justified. Kampmeier was too sharp to have believed this. He was compelled to this conclusion by the pressure to hold the line on rates.

The other and more widespread mining abuse was in the prevailing method of strip mining. Here, the overburden of dirt, rocks, and coal too poor to sell was moved to one side of the coal seam by large-scale earth-moving equipment. The saleable coal was then removed. Typically, when the seam was exhausted, the men and equipment would be moved to a different location, with no effort to restore the land to its original contours, or to contours that would make the land useful for any purpose. Instead, the mounds of spoil would be left in place to erode and clog the streams; the excavated areas would become treacherous sulfur ponds or bogs. The conservation debate over strip mining was fueled, in part, by the increasing share of the market served by such mines. New excavation equipment of great size and efficiency made it possible to dig deeper seams than heretofore. At the same time, they created higher ridges of spoil material and larger craters. The TVA, as well as other utilities, had a significant financial stake in the maintenance of a low-cost strip mining industry, not only for the direct savings in purchases from such mines, but also because of the pressure exerted on coal prices generally.

The clamor for reform was great, but for years little was done. The federal government insisted that this was a problem for the states. The coal operators said there was no need for further legislation because the states already had good laws on the books. The trouble was that the state laws, in the main, were not enforced. The state enforcement agencies had neither the manpower nor the administrative and technical skills needed for effective enforcement. Poorly paid inspectors were frequently corruptible, and the laws themselves were not nearly as well drafted as the coal operators claimed.

The TVA might have imposed obligations on its coal suppliers to mine in such a way as to minimize damage to the land, and to restore it for agricultural purposes afterward. It demurred, however, on the basis that the regulation of strip mining was appropriately a legislative responsibility, and that it did not have the resources for enforcement. A question of the TVA's authority to impose such conditions in its contracts was also involved. I was of the view, which I expressed at more than one board meeting, that the TVA was not compelled to buy coal from mines that did not observe reasonable conservation practices, but I was not asked for a formal opinion.

Air pollution was another serious and growing TVA problem. Much of the coal that the TVA used in its steam plants had a high sulfur content. During combustion, most of it was released to the atmosphere through the power plant stacks as sulfuric acid and other sulfur compounds. Another harmful group of emission products was nitric acid and other nitrogen compounds known collectively as "NOX." In the days of environmental innocence, the TVA power plants were built with short stacks intended primarily to assure that down-drafts would not deposit soot and gaseous effluents in the immediate plant vicinity. Dilution of the harmful effluents by dispersion in the atmosphere was thought to be adequate protection for more remote areas. The early plants were built before the time that electrostatic precipitators had been developed, so that collection of fly ash and other particulates was inefficient by modern standards. No technology had yet been developed for reducing sulfuric acid and NOX concentrations in the stack gases.

The risk of spending decisions that, in retrospect, could turn out to be wrong, is a luxury for the affluent, whether of individuals, countries, or regions. The Tennessee Valley area was only midway in the climb from impoverishment to an equal share in the nation's wealth. Low electric rates were regarded as the principal engine of economic progress. The environmental costs associated with mining and burning various fuels were considered part of the electric business. Only later did a national consensus emerge that stressed the importance of environmental protection. The TVA's environmental program was a story of too little, too late. During my tenure, it was the weakest area of performance of TVA management.

Lilienthal's Idea for a Privatized TVA Power System

In his journal entry for July 20, 1951, Lilienthal reports that the previous evening he had dinner with me ("as long a professional friendship as I have"). He also indicates that we spoke about his refinancing idea for the TVA: "I judge I must have made quite an impression on Gordon [Clapp, then TVA chairman] about it, and that he has been working on Swidler, trying to get him to take a more optimistic view of the idea." This was an understatement. In fact, I was placed under pressure not only to approve the idea, but also to find a way to put it into effect. I would have enjoyed the dinner more if the conversation had avoided this topic. Lilienthal's "refinancing idea" involved a great deal more than refinancing. His plan was to divorce the TVA power program from dependence on Congress or the president by a congressionally authorized transfer of the power properties to a nonprofit corporation headed by a board made up of representatives from the seven states in the TVA power service area. The details of the idea were not filled in; they were left to the lawyers and the power staff. What Lilienthal wanted, and what Clapp soon demanded, was a workable concept on which the idea could be fleshed out.[1]

Lilienthal had resigned the chairmanship of the AEC in February 1950. Within a few months he formed an affiliation with Lazard Freres, a well-known international banking house, headed by André Meyer, a legendary figure in financial circles. Lilienthal greatly admired him and the two became warm friends. The TVA refinancing idea was an early outgrowth of that relationship.[2]

There were several reasons the TVA needed greater autonomy. The most pressing was the uncertainty of securing large appropriations from Congress year after year for funding the financial needs of a vast and rapidly growing power system. The broad, bond-issuing authority for power purposes lay some years in the future, and each year the appropriations route was plagued with hazards that threatened the outcome. Clearly, the TVA could not safely carry out a power supply mission on a solid basis without a more secure source of funds.

As reflected in the comment quoted above, the key TVA staff members, myself included, were not persuaded [by Lilienthal's idea]. We had some good reasons. These could not be strenuously argued beyond a point because the decisions were not ours to make. Having presented our case, without effect upon Clapp or Lilienthal—who were unenthusiastically supported by Harry A. Curtis and James P. Pope—we cooperated fully in attempting to surmount the difficulties the proposal involved. Lilienthal and Clapp seemed to think I did not try hard enough. They suggested that if I had devoted the same intensity of effort to disposing of the TVA power system that I had exercised in helping to put it together, I would have found a way. I did what I could to assuage their suspicions, but not very successfully. They were wrong if they believed I dragged my feet to preserve my job or the accustomed TVA management modes.[3]

My principal difficulties with the proposal, which I believe were shared by the power system leadership of Wessenauer, Kampmeier and Seymour, were as follows: (1) Any administrative scheme involving seven states was certain to be cumbersome and to incorporate centrifugal forces; (2) The TVA was much more than a power project, and each of its programs gained strength from the others. A transfer to state or nonprofit auspices would involve the dissolution of the unified program, with the federal government retaining authority over the navigation, flood control, fertilizer, forestry, land use, munitions research, and recreational programs, which yielded no income; (3) The financing problem was serious, but it could be resolved in other ways (as it proved to be with the passage of the bond financing amendments to the TVA Act); (4) The timing was risky. Once the TVA surrendered on the principle of unified administration, Congress would work its will on any bill, and what came out might have little resemblance to what went in; and (5) Without legislation in each state waiving taxation, any cooperative or other private agency would be subject to taxation at will.

These and other points were made and quickly rejected. The discussions reverted to the feasibility of carrying out Clapp's directive, and here we were met by hard obstacles I could not see my way through or around.

The instructions I received were vague. Create an agency, I was told, in the private sector, if possible, that would carry out the power program on a self-financing and independent basis. It would also have to be able to issue bonds, not only for new construction, but also to repay the federal investment, and in a way that did not compromise the remainder of the TVA program. Very early in the discussions with Clapp and Lilienthal, perhaps at the meeting when the project was first broached, I said that a clear-cut way to carry it out was through a compact among the states in the TVA power area. The drafters of the U.S. Constitution, in article 1, section 10, paragraph 3, prohibited any state, without the consent of Congress, from entering "into any agreement or compact with another state, or with a foreign power." This is the so-called "compact clause," under which many multi-state organizations have been created, each with the approval of Congress.

Clapp and Lilienthal quickly vetoed the compact approach on the basis of time and difficulty of persuading seven state legislatures and governors to agree on a common plan. As time passed and the problems with other schemes became clearer, I again brought up the idea of a compact. Perhaps, I suggested, a compact could be started with an initial group of three or four states, if these included Tennessee and Alabama, the states with the largest shares of investment and markets, and if the door were held open for the others. After all, the framers had provided that the Constitution should go into effect upon the approval of nine of the thirteen states. I pointed out all the advantages of a compact-based agency. It could combine both federal and state powers, and eliminate the shadowy area of the relative reach of state and federal jurisdiction. Once established, it had great institutional stability because the compact could only be changed in the manner provided therein, and only with the consent of Congress. But Clapp and Lilienthal were not convinced. Pass a miracle, they said.

I assigned several lawyers to the investigation of the legal consequences of creating a multi-state agency without benefit of a compact. It became clear that it would become subject to taxation and regulation. Exemptions in one state would have no effect in others. Without a compact there could be no binding agreement on structure or mode of appointments to the governing body. Not even the right to issue tax-exempt bonds would be available. It would be easier to domesticate a herd of hippogriffs than to create a public/private agency meeting the Lilienthal-Clapp requirements.

Not satisfied that the TVA lawyers had exhausted all the possibilities, Clapp and Lilienthal suggested that I should retain Joseph Volpe, a Washington lawyer, as a consultant. I was pleased with the suggestion and promptly engaged Volpe. He was well known to me, an excellent lawyer of independent mind who had served as the AEC's general counsel during a part of Lilienthal's tenure as chairman. I showed him all the legal studies we had made, and offered to investigate any other ideas he might have, but nothing came of his participation. If he saw daylight on the Lilienthal plan, he never mentioned it to me. I welcomed him because I thought he would lay to rest any suspicion that I or the other TVA lawyers had failed in diligence or imagination. He must have done so because the "refinancing plan" was quietly abandoned after another month or two. It was not particularly gratifying that Clapp thought it necessary to bring in another lawyer to check up on me; he was entitled to a second opinion, though, and Lilienthal was pressing him.[4]

Having satisfied the reader of the intensity of my effort to find a way to carry out the Lilienthal-Clapp plan, I hope I do not disturb that faith by confessing I thought Lilienthal suffered from a conflict of interest. His assignment for Lazard Freres was to find new scope for the firm's financing activities; refinancing the U.S. Treasury's investment in the TVA would have been a profitable coup for him and the firm. He should not have undertaken this mission, and except for the opportunity for potential financial benefits and to establish himself in Lazard's entrepreneurial world, he might well have agreed with me that such a situation was not healthy. While I do not believe he had a conflict in the legal sense, Lilienthal was identified with the TVA. He purported to be animated by the TVA's concerns, and had special entrée because of his past affiliation. I do not believe he would have encouraged or sanctioned an arrangement he did not consider to be in the TVA's interest, and the TVA indeed suffered from its dependence on appropriations. His involvement as a friend of the TVA, however, was inconsistent with his normal skepticism about the ability to serve two masters. I never said this to him or to Clapp, which I now regret. Had the plan proceeded further, Lilienthal (and Clapp) might have suffered serious embarrassment. Perhaps Joe Volpe straightened them out. At a later stage in my career, when I was more sure of myself, I am confident I would have warned Lilienthal and Clapp. I would have also distanced myself from the proposal and the TVA if they had not heeded the warning.

CHAPTER 11

The Dixon-Yates Scheme, 1953–1955

One of the great scandals of the Eisenhower administration was the so-called Dixon-Yates scheme. Eisenhower believed the TVA should be privatized, although he managed to avoid an explicit statement to that effect during the 1952 campaign. In 1953, however, with some assistance from the First Boston Corporation, the Bureau of the Budget concocted a plan to stop the TVA's growth. If the plan had been successful, it would probably have led to the early dissolution of the TVA and the demise of its power program. The scheme failed, a victim of its own conspiratorial origins.[1]

The background is quickly sketched. The TVA had been growing so fast that it outran its ability to raise money for new generating projects. It had no bonding capacity for the purpose; net revenues alone were inadequate, and Congress was progressively more resistant in granting new appropriations. The logical place to build the next generating plant was near Memphis, which was supplied over long transmission lines extending westward from the main TVA system. In terms of system planning, Memphis required a local generating source. The TVA proposed such a plant, named Fulton, to be located on the Mississippi River at a Tennessee site above Memphis, but Congress failed to supply the money. At the time, the AEC's operations at the Oak Ridge and Paducah plants, plus a few other federal installations, required half of the TVA power supply. To conserve its power resources for its native area loads, the TVA requested that the AEC eventually reduce its take by 500,000 kW. The Bureau of the Budget thought it had a better idea—Dixon-Yates, as it came to be called.

Edgar H. Dixon and Eugene A. Yates were the respective heads of the Middle South Utilities Company and the Southern Company. Yates I had come to know very well. The two men were bitter and vengeful. They were fearful of further incursions by the TVA, despite assurance that it had no intention to expand, an assurance backed up by a decade of rejected applications from municipalities and cooperatives outside its defined market area.[2]

The Dixon-Yates scheme was elegant and not overly complicated. The Middle South and Southern Company formed the Mississippi River Generating Company (MRGC) for the purpose of constructing a power plant on the Mississippi in Arkansas. The AEC would then contract with the MRGC for delivery of power (over one or more river crossings) to the TVA in Memphis. In turn, the TVA would transmit this power to the AEC.[3]

The plan did not present an easy target. The physical system would be almost identical with the one the TVA proposed. A private facility [in contrast to the TVA-operated Fulton plant] on the Mississippi near Memphis would generate power at one end, convey that power to the TVA in Memphis, which would then deliver it (by transmission relay) to the AEC on the other end. The TVA would lose the AEC load, but it had asked to be relieved of it. In fact, I had argued with Chairman Clapp that we should not ask the AEC to reduce its reliance on TVA power. I did not anticipate the Dixon-Yates intrusion, but believed the action was premature and that the TVA's best course was to build new and even more efficient stations rather than withdraw within the shell of its existing system and survive by forcing the federal government [that is, the AEC] to look elsewhere for power suppliers. Clapp was adamant. I never fully understood his reasons, but I could not press him. The other members of the board did not do so either; they may very well have been content with Clapp's strategy. The only substantial differences were that the private companies would now have a customer in the heart of the TVA area, and more important, as a physical matter, they would be supplying the Memphis load, and Memphis would be dependent upon them.[4]

I first learned of the scheme one morning when I found on my desk a request from the general manager to prepare an answer to a letter he attached, from the White House, for the signature of the chairman. The general counsel received many such requests, and much of the time of several lawyers was devoted to the preparation of draft responses. The form provided space for directions as to the nature of the response, or of special problems to be considered. This time, however, no instructions or suggestions were given.

The letter was not a run-of-the-mill White House document, but was signed by the president himself. It outlined the plan and instructed TVA to cooperate by executing the necessary contracts. As I recall, it stated that the AEC was being given corresponding instructions.

This was a tough letter to answer. That Eisenhower had signed the letter himself, rather than having it signed by someone else "at the direction of the president," gave it special emphasis. I do not recall another such letter. It was intended to eliminate all choice by the TVA. To defy the president was impossible. Equally obvious was that an affirmative response might result in the destruction of the TVA.

After much thought, I followed the bureaucratic rule: When in doubt, stall, especially if time is on your side. Accordingly, I told the president that he could not have been fully informed; I then proceeded, in the most persuasive way I could manage, to make the TVA case against conforming, requesting the president to reconsider his decision. I did not ask him to hurry a reply. I knew the TVA response would pass through several hands before it reached the person who would provide a first draft of the answer, and that several people would make changes as the answer wound its way back to the president. No reply could be expected for several weeks. In the meantime, the news would spread and the TVA's friends would be alerted. The letter as drafted was signed by the chairman and sent to the president.

It appears, although I did not know it at the time, that three of the five AEC commissioners opposed the Dixon-Yates plan—Murray, Smyth, and Zuckert. In the end, however, the AEC was forced into line, and the head of its Oak Ridge office notified us that he was ready to negotiate the necessary contracts to put the plan into effect. He was a good friend of the TVA, but also a responsible government official who did well the work assigned to him. In time, the president reiterated his directive to the TVA to begin the process (in effect) of dismantling itself. The TVA had no choice but to comply, and a long and tedious negotiation with the AEC ensued.[5]

I wrote Yates a "Dear Gene" letter, referring to the stabilization agreement reached between Lilienthal and Willkie. I questioned the wisdom, and perhaps the good faith, of the attack on the integrity of the TVA power service area. I asked whether the scheme was in the long-run interests of the utilities. I received a rather stiff reply in which he said his companies were always interested in expanding their businesses. He saw no barrier to the current project.

While the TVA and its officers and employees were constrained in their opposition, and indeed compelled to cooperate, a tumult of protest arose from outside interests. The people of the TVA area were appalled, and their state and local representatives raised several objections. In Congress, the proposal was fought with growing intensity and success. The individual distributors and the state and national trade associations of the municipalities and cooperatives were active in fighting the plan. Of crucial importance were two local initiatives. One was by the city of Memphis, which announced that under no circumstances would it receive power from the proposed Dixon-Yates plant, and that to protect itself it would terminate its power supply contract and build its own power plant in the vicinity of Memphis. Memphis made good on its threat and built a large plant named after the longtime head of its municipal utility system, Thomas H. Allen. There is a suggestion in Lilienthal's *Journals* that the idea for this initiative may have come from Lilienthal, who reported in an entry for September 26, 1954, that he had phoned Allen to say: "The Valley has strong alternatives; they don't need to be kicked around this way." The other [initiative] was the retention of Joseph Volpe as special counsel by the Tennessee Valley Public Power Association, the trade group for all the TVA distributors. It was Volpe who helped bring this doomed adventure to an end. He unearthed the fact that First Boston had placed one of its men, Adolph H. Wenzell, in the Bureau of the Budget, as a dollar-a-year man, to work out the Dixon-Yates program. The news, when it came out, created a sensation. Opposition mounted fast. Memphis gave the plan the coup de grace, and, in July of 1955, Eisenhower formally canceled his directives. In 1959, Congress found a better way to finance the TVA—by the issuance of revenue bonds—not obligations of the United States, but payable only from power revenues, and carrying no exemption from federal income taxes for bondholders.[6]

CHAPTER 12

Fired by General Vogel, 1957

In late summer of 1957, General Herbert D. Vogel, who had been appointed as chairman of the TVA by Eisenhower three years earlier, was the only sitting member of the board. The term of Harry Curtis had expired in May of 1957, and Raymond Paty died of a heart attack in August of 1957. The General was in sole charge. Within a few weeks, he asked me to resign.

Until Curtis's term ran out, General Vogel had been a minority member. Curtis and Paty, both serving in the New Deal tradition, voted as a bloc. A quiet bitterness developed in Vogel as he was overruled again and again. He bided his time, awaiting appointment of another director to replace Curtis.

After Curtis left the board, there was a stalemate for a brief period. The Senate postponed action on an Eisenhower nominee, Arnold R. Jones, since Democrats suspected that, if confirmed, Jones and Vogel would dismantle the TVA. This suspicion was widely shared by the TVA staff, including me. With Paty on board, each [Vogel and Paty] could veto an initiative of the other, and the TVA administration settled into a holding pattern. With Paty's death, the situation was transformed. A quorum of two members was required for certain actions, but Vogel now had a free hand on many personnel matters. His authority on internal questions was unchallenged, especially since it was clear that Eisenhower would ultimately have his way in filling the two vacancies.[1]

General Vogel's request for my resignation came in September of 1957. I described the triggering circumstances a few years later in my testimony before the Senate Committee on Interstate and Foreign Commerce, which was considering my nomination by President Kennedy to serve as a member and chairman of the Federal Power Commission. I can do no better here than to quote from the official hearings. The questioners in this excerpt were Senator Scott, a Republican and ranking minority member of the Committee, and Senator Schoeppel, also a Republican:[2]

> SENATOR SCOTT. I have been receiving newspapers from Tennessee and these papers evidently are being sent to me because they want me to clear up

one thing. The *Knoxville Journal*, September 13, 1957, has a story which begins: "Swidler Resigns as TVA Counsel."

And then on Saturday, January 28, 1961, there is another statement: "Swidler"—well, the caption is "Kennedy's Selection for FPC Chairman, Swidler, Once Fired, as TVA Legal Chief." And the statement says: "Efforts to block the Jones appointment have brought the discharge." The news article says in part: "The reason for his discharge, cloaked as a resignation at the time, was that Swidler preferred being fired quietly to resigning on request, thus he could protect his civil service retirement benefits. His resignation had been asked because TVA chairman Herbert T. Vogel discovered Swidler was using agency personnel in preparing loaded questions intended to give Arnold R. Jones a hard time during Senate committee hearings on his appointment to the board."

I am asking you to comment for the purpose of clearing this up. I want you to know it is not any personal hostility on my part. I rather think that when the time comes, there is unlikely to be any opposition from me. I would like you to know I am not influenced by planted stories in morning papers either way, but I do think that in fairness to the commission and to the Congress, that you ought to be given an opportunity to clear this up, since the newspapers in your home state appear to be making a great thing of it. Would you care to comment as to what the actual facts are?

MR. SWIDLER. I would be glad to tell the whole story and I must say that I appreciate the very fair and neutral way in which you introduced it.

I might also say I was rather flattered at the attention that the *Knoxville Journal* gave the circumstances of my leaving TVA several years ago. At the time this story appeared there were big things happening in Laos, and troubles in the Congo, events I thought had far transcended this matter, but they thought this was the most important news of the day and gave me streamer headlines.

SENATOR SCOTT. They have been sending me copies of it. I therefore feel obliged to bring it to your attention.

MR. SWIDLER. I think perhaps the question you ask involved several different matters and I will, if you like, start from the beginning and tell you the story in a way that may explain them all and if it doesn't, why you can ask me questions.

SENATOR SCOTT. Handle it in any way you see fit so that you may have a fair opportunity to comment on them.

MR. SWIDLER. As you know, there was a change in administration in 1953, and that was reflected in changes in the TVA board. And in the period from 1954 to 1957, there was a complete change in the complexion of the board. This was a difficult period for the board and I might say for staff as well. It was inevitable that some tension, some misgiving, should have been created in this period.

During, as I recall, spring and early summer, there were hearings or proposed hearings on the nomination of Mr. A. R. Jones of Kansas as a member of the board. At one stage in connection with discussion of his appointment, the TVA Washington office handed me a thick document relating to the development of the rivers of Kansas and I was asked to analyze it—I think it related to the natural resources of Kansas, rivers of Kansas.

SENATOR SCHOEPPEL. With your indulgence, might I ask you, who gave you what in this case?

MR. SWIDLER. It was given to me by the TVA's Washington representative, Miss Margaret *[sic]* Owen, and it was a volume the name of which I no longer recall, but it was prepared in Kansas and dealt with the development of Kansas's natural resources. This is about as much of it as I can remember. It was too big and thick for me to go through myself at the time. I turned it over to one of the young men in the office and I asked him—this was not transmitting the request I received, but going beyond that, and on my own initiative, I asked him to prepare some questions which would bring out the views of Mr. Jones, if submitted to him, on the policy questions with which he would be faced as a member of the TVA board.

This young man prepared a set of questions in response to my request and I must say that they were not a very good set of questions. They were never approved by me nor were they used.

As I recall, shortly after this time, Dr. Paty, the last holdover of the previous administration, died, I think, in August. Dr. Curtis, who had been the third man on the board, his term expired in May. In September, a copy of this set of questions was apparently brought to the chairman of the board, General Vogel. He was at that time the only member of the board. Mr. Jones, I think, had been approved but had not yet been sworn in. General Vogel called me down to his office. I have refreshed my recollection on the date, it was on September 11, 1957. He called me down to his office and told me that he had acquired this set of questions, that it had impaired his confidence in me to the extent that he thought I could not serve usefully as general counsel of TVA, and he asked for my immediate resignation.

I arranged to see him the next day, being too shocked at the time to respond immediately. When I saw him the next day, I told him that I had no question as to his right to have anyone that he wanted as his general counsel and that if he, as I say, then being the sole board member, if he wanted me to leave for this reason or any other, or no reason at all, I would go.

At the same meeting I told him that under the civil service regulations in force at the time I was entitled to a pension, that someone who had reached the age of fifty, as I had, and who had more than twenty years of service, as I had, and whose separation was involuntary, as this certainly was, was entitled to a pension. And I asked that the record be made clear that the termination was involuntary so that there would be no mischance on the entitlement to a pension. He agreed to that and to other details of the arrangements. It was a fairly amiable discussion the second day when he found I would not insist upon staying as general counsel when he wanted me to leave. And, accordingly, thirty days later, I was terminated and that concludes this little story.

Now, have I left anything out, Mr. Scott? If so, I will be glad to come back to it.

SENATOR SCOTT. Nothing that I know of. I simply raised the contrast between the two news stories to give you a chance to explain it, that is all.[3]

The Committee promptly recommended approval of the nomination, and the Senate gave its approval (albeit with some delay). Only Senator Strom Thurmond of South Carolina voted in the negative. Senator Thurmond had questioned me closely on the fairness of government subsidies to public power systems, and apparently was not satisfied with my responses. Later, when I represented a South Carolina agency, I found the senator and his staff easy to work with, and relationships were cordial. To my surprise, I discovered much to admire in the senator.[4]

My termination received wide notice in the press, which also reported several comments. One was by Justice Felix Frankfurter, who said that anyone in a staff position who let himself get involved in policy should expect no better. Coming from Frankfurter, who was notorious for his efforts to influence policy from whatever position he might occupy, including that of a Supreme Court Justice, this seemed an unkind cut. It came at a time when a friendly statement would have been much appreciated, but on the merits he was right.[5]

CHAPTER 13

A TVA Balance Sheet

In those heady days of large plans and great achievements, it was easy for the members of the board to take literally the bold rhetoric of the president in his 1933 message to Congress. Lilienthal's 1944 book, *TVA: Democracy on the March,* and the book by his protégé and successor, Gordon Clapp, written a decade later, are panegyrics to the wisdom, if not the clairvoyance, of the people who wrote the TVA Act—and inferential tributes to their own administrative talents.[1]

If the TVA was the product of such profound wisdom and magnificent administration, why has no similar authority been created elsewhere in the United States? Many people here and abroad have asked me this question. Frequently, it was clear that the questioner thought he already knew the answer. The wicked U.S. power trust was so omnipotent that it was able to kill off any effort in Congress to duplicate the TVA's success. True, the utilities strongly opposed the creation of any other organization on the TVA model, but this was a minor factor. The political influence of the utilities has been waning steadily for several reasons—and that may deserve a book of its own. No, the primary reason was that the TVA Act was written to fit the unique circumstances of the Tennessee Valley area at a particular time. The combination of great poverty and great opportunity, centering on the neglected resources of a major river basin, did not exist elsewhere.

The difficulty in finding an integrating factor or principle for the establishment of regional agencies with broad natural resource functions was brought home to me in the mid-1930s, when, at the request of the White House (via Benjamin V. Cohen, Roosevelt's premier legislative draftsman), I was asked to draft a bill to establish regional agencies throughout the country. Working with Cohen's brilliant and devoted young assistant, Henry Herman, we developed such a draft that was introduced in both houses of Congress. It was the subject of extensive congressional hearings and a great deal of public attention. The familiar names of the bill were Norris-Rankin, after its Senate and House sponsors, and the "Nine TVAs" bill, its descriptive name.[2]

An early task in writing the bill was to define the respective regions. It was not difficult to define watersheds, and ultimately we did so, without benefit of technical advice, I regret to say. Three of the regions had already been delineated in prior legislation or by executive action. The TVA, of course, was well established. The Bonneville Administration, as a power-marketing agency for electricity generated by projects on the Columbia River and its tributaries, was another extant region; one of the purposes of the bill was to broaden its charter on the TVA model. Because of expected opposition and delay, I was eventually asked to draft a simple interim bill to tide Bonneville over for a year or two until Norris-Rankin was passed. The third area that was the subject of prior definition was the Arkansas River Valley. A subsequent bill [to establish an Arkansas Valley Authority] was parented by Representative Clyde Ellis, a strong public power partisan who became one of the early heads of the National Rural Electric Cooperative Association (NRECA). Even with this start, Herman and I still had most of the country to divide into regions. We worked in the Department of the Interior building in Washington with the aid of a big map. In the course of a day or two, we disposed of this chore and secured Cohen's approval of our hasty map work, which was probably similar to, and no worse than, that of the victors of World War I at Versailles, with the unfortunate difference that the 1918 results led to another world war. So far as I know, there were no tragic consequences from the project completed by Herman and me.[3]

We followed our mandate and drew boundaries around nine areas of the country, based on river basin lines, but the project had become a game. With the exceptions I have mentioned, there was no focus within the respective regions. Even in the TVA, we had found that a hydrologic area was not an effective unit of administration for any purpose except water control. The TVA's power service area, for example, was twice the size of the Tennessee Valley itself. The agricultural problems with the Tennessee Valley were not necessarily different on one side of the watershed than on the other, and the TVA's fertilizer program was soon extended to a national level, with test demonstration farms in almost every state.[4]

In part, the TVA was designed to correct a regional imbalance in productivity, personal income, and standard of living. These, too, by their nature could be applied only selectively. The TVA started with a legacy of hydroelectric, steam, and fertilizer plants that had been built for World War I purposes. This combination made possible the use of these facilities as a center for fertilizer research and for farm demonstrations. Ultimately, they were the basis on which the TVA fertilizer plants and laboratories at Muscle Shoals became the national fertilizer center. There could not be nine such centers. By its inherent nature, this leg of the TVA program was unique.

An even more fundamental question was raised by the Norris-Rankin proposal to nationalize the TVA concept. The TVA in effect organized certain powers of government on a horizontal basis. It was based on integrating all national resource functions in a defined area. In the rest of the country, however, the administration of these functions is configured on the basis of vertical organization. Some agencies build dams, others operate them. Some operate and administer parks, others deal with forests. Agricultural programs are a specialty of one department. Navigation is entrusted to a different agency than the one that markets the power from a dam. The TVA, in one corner of the country, could be accommodated within existing government structures because of its functional specialization; but, if TVAs had been created throughout the country, many bureaus and departments would have lost much, if not all, of their jurisdiction. The country was prepared for an *experiment* in regional administration, but not for the fragmentation of national agencies that controlled resource development. Needless to say, the agencies themselves resisted and resented the TVA. Their opposition was far more effective than that of the utilities. Basically, it was Congress itself that retreated from the implications of regional authorities on a national scale.

Another problem was that watershed lines generally did not correspond with existing concepts of regionality. New England is, or was, a distinctive region in most minds, though it was not defined by watersheds. The same was true for most coastal areas. In other cases, a river had a large drainage area —the Mississippi or the Missouri, for example—but the various parts of the drainage area had little in common. In some cases where size alone was not an obstacle, there was neither hydropower potential nor the possibility of large improvements in river navigation. Sometimes regionality was expressed in agricultural terms—the Cotton Belt or the Corn Belt—and in others by mountains or lakes, by deserts, or by water-use regimes (reliance on irrigation, specifically), yet river boundaries rarely provided a common cultural bond or way of life. Finally, the implications of a national system for regional administration of resource programs became apparent. Regionalization of federal authority throughout the country would have involved radical changes for which no strong case was made. It is no wonder that support for Norris-Rankin dwindled, that it finally dropped from the political scene, and that the TVA remains unique to this day.

A Personal View of David Lilienthal

Lilienthal is a historical figure. His portrait now hangs in the National Portrait Gallery, a fitting recognition of his stewardship of the TVA and the AEC, and of his impact on American life. He was my first employer when I became a lawyer in 1930. In the years of his TVA service, I knew him as well as anyone, his family aside. Afterward, we met from time to time, occasionally on business matters. I kept in touch, directly and through friends, and also through the press and his writings. He was a man of many conflicts, although this side of him rarely showed. He had a deep influence on the course of my life, and whatever happened to him was of interest to me. I write of him now, partly to put my own views in order, and also because he had a remarkable and varied career. He left a unique record in his *Journals* of American politics and business in the middle decades of the twentieth century.[5]

Any Lilienthal biographer must start with Lilienthal's *Journals*, which cover several thousand pages in seven large volumes. Few contemporaneous sources provide better insights on the sweep of events in government, politics, and business, and the relationships among them. At the same time, if the reader is not to be misled, he should keep in mind some facts about the *Journals*. Most important, they were written with eventual publication in mind. On personal matters they cannot be accepted as candid. He was under no obligation to mention personal matters at all, but he does, and—with a few notable exceptions—they always contain good news, or news that would reflect well on him and his family. The *Journals* are not a record of the life Lilienthal would have lived if he did not keep a journal. In the camera's eye he trained on himself, he tended within the limits permitted by his career to live the life that would reflect well in the *Journals*.[6]

Lilienthal's mind was usually in high gear. He had quick and intelligent opinions on much of what he saw and heard. He was a decisive administrator, an astute political observer, and although not a partisan politician, he could swim safely even in troubled political waters. His critical views were kept under reasonable control. In public, including his *Journals*, he spoke well of most of the people he met, but he did not shirk from frank and sometimes pungent appraisals of people he did not like. Even here, though, his comments gave little clue to the acerbity of some of his private appraisals. He was troubled, as ever, with the conflict between his desire to serve the public good and the attractions of wealth and prestige.

Finally, despite his sophistication and worldliness, he remained at heart the product of the small town of his youth. In his 1971 description of a board meeting of the International Basic Economy Corporation, he said that the presiding officer was Dick Dilworth, "alert, handsome, with the best manners of anyone I know (and don't let anyone tell you manners aren't a mark of a man's insides)." Lilienthal's emphasis on manners as the key to character throws light on many of his appraisals of people and his attitudes toward them.[7]

Not all [TVA] achievements were due solely or primarily to Lilienthal's efforts. A.E. played the lead role in dam building and in planning for river management, and H.A. in the fertilizer and farm improvement programs.

Lilienthal, however, could take the principal credit for acquisition of an electricity market in the face of tremendous opposition from the utilities that served the area, and despite the fact that A.E. supported the utilities and their formidable leader, Wendell Willkie. It was Lilienthal who insisted, and finally prevailed, in marketing the TVA's power output primarily through public and cooperative agencies, agencies that agreed to pass on to consumers the benefits of the federal investment and enterprise. It was Lilienthal, also, who must be given credit for assuring White House and congressional support, for waging battles for internal independence against voracious central agencies, such as the General Accounting Office (GAO), and for opposing Secretary Ickes's endless machinations to swallow the TVA.

That Lilienthal proved to be a charismatic leader was a surprise to him, and his success was all the more remarkable because he was a northerner in the Deep South, a federal administrator in a states-rights political environment (one of his strong supporters was named States Rights Finley, a South Carolinian transported to Tennessee), and a Jew in the Bible Belt. To be fair, the TVA program did not ask for sacrifice, but instead promised to save money for consumers, to create more jobs, and to raise the standard of living. Still, even grant programs could run into trouble, and it was clear that Lilienthal's popularity and persuasiveness were due as much to his personal appeal as to the inherent attractions of the program. As for his being a Jew, this was not a feature that he emphasized. He belonged to no Jewish organizations, and, during his TVA years, he and his wife, who was not Jewish, attended a nondenominational church in Norris.

During the whole of his TVA tenure, as reflected several times in the first volume of his *Journals,* he abjured wealth as his destiny or as the object of his career. He said the same thing to me numerous times in our talks. I believe he was sincere, although from the laboring of this question it might well be deduced that he possessed some inner conflicts. Perhaps it should be added that, so far as I knew, no great temptations had tested his resolve to devote his life to public service. True, Willkie had made expansive references to future rewards if Lilienthal, with his magnetism and managerial skills, should begin to see merit in the utility position, or so Lilienthal told me, not necessarily in the words I recall a half-century later. Such a suggestion was crude and dishonorable, and Lilienthal referred to it with contempt. It is hard to reconcile the incident with the image Willkie projected some years later as a thoughtful and liberal Republican candidate for president, but Willkie was a hard fighter. At the time of his battle with the TVA, he was a young man in a hurry, and probably reflected the norms of his time for a businessman in his position.

Lilienthal was sophisticated enough to know the career benefits of keeping himself in the public eye and to present himself in a favorable light. The conventional way was to write a book, and perhaps the publication in 1944 of *TVA: Democracy on the March* was an early indication that he was casting around for new employment ventures. The book was a great success. It was lofted by strikingly strong praise from, among others, Vice-President Henry A. Wallace, Charles A. Beard, and Stuart Chase, the latter two among the most influential expositors and critics of social and political trends. Favorable reviews sell books even when, and sometimes because, the reviews bear little relation to the books themselves.[8]

Despite its title, the book is not primarily about the TVA. As Lilienthal states in the preface: "This book does not purport to be a full account of the TVA. . . . What I have set down in this book is a statement of faith, and the facts and reasons that support it."[9]

Lilienthal's reliance on faith should have been a warning, because on secular matters, at least, faith is a weak crutch. It is too bad that Lilienthal did not see fit to tell the TVA story, especially on the power controversies in which TVA was engulfed, because he had more to contribute to history on this turbulent and exciting part of the TVA's history than anyone else. As he acknowledges with pride at several places in his *Journals*, his greatest talent was in making people believe that what he was doing was of major importance, that his experience demonstrated universal principles. In my view, the book stretched the TVA experience a good deal to support his generalizations. It has done harm by raising unrealistic expectations and lending itself to sinister explanations of this country's failure to follow up on the TVA experience by creating similar agencies in the rest of the country. At the same time, the TVA concept as described by Lilienthal became a call to action in many developing countries, a source of idealism and hope. Lilienthal could not have wished for a better or wider response.

Lilienthal referred parts of his manuscript to me for review and comment. I tried to be candid and helpful. Some of my suggestions he adopted, but he had a natural difficulty with comments critical of the manuscript's central theme. I told him the book misconceived the basic nature of the TVA and overstated both its importance as an exemplar of improved government administration and the benefits it had conferred on the Tennessee Valley area. A few of the more extreme statements he tempered a bit. The manuscript, however, was not significantly changed on the questions I had raised.

A few examples show the overblown nature of the book's claims. Chapter 14 begins: "What I have been describing is the way by which in this region we are working towards a decentralized administration of the functions of the central government." Had it been accurate, this characterization would have given the book a lasting importance. But the TVA did not displace a single agency or activity of the central government. It was a modest overlay.[10]

Another example of Lilienthal's tendency to confer on the TVA a stature exceeding its statutory scope is his description of it as an "autonomous" agency—and the statement that such autonomy was the key to a generic category of other decentralized regional resource agencies modeled on the TVA. The basis for this reference was that the TVA was organized as a corporation, was not part of a government department, had its own personnel system, and was partially exempted from the GAO's controls. These factors were critical in helping the TVA perform its responsibilities, but they conferred neither substantive authority nor freedom from strict scrutiny by the Bureau of the Budget and congressional appropriations committees.[11]

I have mentioned elsewhere that when, in the mid-1930s, I prepared for the White House a draft of Norris-Rankin, I could find no solid basis anywhere for a regional agency like the TVA, except for the Columbia River basin in the Pacific Northwest. I reminded Lilienthal of this experience, and he attempted to deal with it in chapter 15. Wisely, he did not try to define a region; instead, he referred interested readers to the "substantial literature" on the subject. He acknowledged that river basins were not necessarily the basis for defining a region, but said that the TVA form could readily be adapted to other resource problems. He said regional consciousness already existed elsewhere in the country, as indicated in phrases such as, "'Senators from the Corn Belt'" or the "'cotton bloc,'" or the "'New England delegation to Congress.'" The views expressed by these examples, however, would not necessarily support the larger concept of an autonomous regional administration. People think of themselves as citizens of a state, county, town, or city, not of a region.[12]

Lilienthal's claims that the TVA discharged its river management, electric power, and fertilizer responsibilities in a successful manner were well taken and skillfully explained. He made a persuasive case for administrative delegations and for strengthening the capabilities of local governments. But he went far beyond these matters in his claims for the TVA as an experiment in regional administration of federal powers. All other federal agencies with resource responsibilities in the Tennessee Valley were alive and well, exercising their traditional roles. Lilienthal's exaggerations were not intended to deceive. He was carried away by a roseate, though mistaken, vision of the TVA's powers, largely because he wanted to impress a national audience with the overriding importance of his work and achievements. In sum, the book's title was a pretentious and misleading characterization of the TVA program, but it also helped capture the imagination and stimulated many efforts at administrative improvement.

At the AEC, Lilienthal became intimate with some of the very rich, including [Bernard] Baruch and Admiral Lewis L. Strauss, both former Wall Street titans. They told him how valuable his talents would be in private business. Strauss (my navy boss, with whom I kept in touch) told me he had suggested to Lilienthal that the two should work as a team in business. With his own technical and financial background and Lilienthal's salesmanship, he believed they would have dramatic financial success. Even before Lilienthal was confirmed for the AEC post, Strauss said to him: "We'll go into business together; I guarantee you'll make a million dollars in three years."[13]

Lilienthal departed the AEC on an unhappy note. President Truman had appointed a "special committee" to submit recommendations on the development of a super, or hydrogen, bomb. Only a few weeks earlier, the Soviet Union had detonated a nuclear weapon, destroying the nuclear monopoly of the United States and the sense of invincibility that the monopoly had fostered. The committee recommended that the nation should proceed with H-bomb research, development, and (by implication) manufacture. All members were agreed, not necessarily for the same reasons, except for Lilienthal, who made an extended statement in opposition. He said that the H-bomb was destabilizing, that it was a sole but impractical and implausible reliance for national safety, that other avenues for assuring the safety of this country should be explored, and that there was no good reason why the report should not be delayed for a couple of months to permit such an exploration. The committee presented its report to the president on January 31, 1950. Lilienthal stated his dissenting view briefly, and the president did not ask him to expand. Truman said the matter was of such intense public interest that he must make and announce a prompt decision (he agreed with the committee's majority). Lilienthal did not make his dissent public, but was pleased to have stood his ground. It seems to me that this was one of Lilienthal's finest hours, whether or not he was correct in his thesis.[14]

Lilienthal's introduction to private business was highly favorable. He started with a contract to write for *Collier's* for substantial compensation. *Collier's* financed Lilienthal's travels abroad, and wherever he went, he was received as a leading political figure. More lucrative speaking engagements were offered than he could accept, and he had a choice among business associations. Lilienthal's first *Collier's* article, published in June 1950, signaled a sharp departure in emphasis, if not substance, from his views on the relative roles of government and business. In the article he "argued against the prevailing policy of government monopoly in non-military and commercial atomic areas." Considering the fears over nuclear proliferation, this was a bold and extreme position for 1950, when the genies were still in the bottle. It was not at all consistent with Lilienthal's cautious approach to nuclear development when he was on the AEC, or his official view on the role of industry.[15]

A year later the change in Lilienthal was made even clearer. Believing that the articles in *Collier's* had not given him sufficient public prominence, he conceived the idea for "a big special issue of *Collier's*" devoted to "'Big Business for a Big Country' theme." Lilienthal would "propose a 'program' to take the place" of the Sherman Act. He would extol the virtues of size and industrial integration, and emphasize the benefits of competition between different products, rather than on competition among sellers of the same product. All this Lilienthal would somehow combine with the "the TVA decentralization theme." He was anxious to please, an attitude hard to avoid in the initial phase of building a place in the business world.[16]

Something of the change in Lilienthal was also reflected in an interview reported in the "Talk of the Town" section of the *New Yorker.* The article reported that "Lilienthal has changed," and that he was happy because he was "making lots of money." Lilienthal questioned the accuracy of the quotations, but not their gist. He and Helen rightly feared that the article would worry their friends throughout the country.[17]

That Lilienthal was a great success as a consultant was clear. He seems to have made many valuable suggestions and to have carried out some activities of financial importance, most of them on the development and exploitation of sources for coal, iron ore, copper, and phosphorous. It was about this time [the mid-1950s], when Lilienthal was riding the crest of a euphoria about his success in business, that he paid a visit to Knoxville and addressed a large group of TVA employees at the S&W Cafeteria on Gay Street. I was present to hear him boast that he had already (in about a year) made his first million dollars. This was not the Lilienthal I thought I knew.

PART V

A Return to Private Practice, 1957–1961

Having been forced to resign from the TVA, Swidler found himself in an embarrassing situation. His case had become a cause célèbre because of the press attention it received, and from the fact that Supreme Court Justice Felix Frankfurter made it the centerpiece of a speech he had given at the annual meeting of the Federal Bar Association in 1957.

In Swidler's opinion, this ambiguous publicity limited his career options. He ruled out working for a utility company since he had "no mind to switch sides in the public power fight." Instead, Swidler served temporarily as a United Nations consultant in New York City, editing a report on river basin development. During that assignment, he had time to reflect on how he could "reconstitute" his "working life." In late 1957, he became a public utility consultant in Knoxville. After a few months there, he and his family moved to Nashville, where he affiliated himself with a small law firm, Denny and Leftwich. For the next three years, Swidler represented a variety of clients, mostly municipal electric systems and electric cooperatives. The pace of work was slower than at the TVA, and the stakes were certainly smaller, yet Swidler adjusted without difficulty.[1]

As Swidler notes, the most important case he handled while in Nashville was the ultimately successful attempt by the city of Paducah, Kentucky to purchase the local electric distribution system from the Kentucky Utilities Corporation. Swidler's description of this struggle illustrates the continuing opposition municipalities faced when they confronted utility companies over the issue of private versus public power (that is, when cities attempted to acquire private distribution systems for the purpose of transmitting TVA power to their citizens). But Swidler's role was more complicated than that of an attorney in a relatively simple legal dispute. In 1959, when he had already begun advising the city of Paducah, he was also a legal consultant to the U.S. House of Representatives Committee on Public Works. In the latter position, he made sure that Paducah was granted an exemption from the territorial limitations Congress placed on the TVA's service area. Though Swidler asserts that his dual role was not a conflict of interest, it was not beyond suspicion. At the very least, it represented the ability of lawyers like Swidler to move between the private and public sector, influencing decisions in both.

Swidler's victory in the Paducah case notwithstanding, he himself voiced some misgivings about whether the outcome was in the "public interest." He believed that the shift from private to public power sometimes represented an unfair transfer of costs. In short, municipalities, unlike private corporations, are able to issue bonds whose interest is free from local, state, and federal taxes. In exchange for this tax benefit, municipal bonds normally yield lower interest rates; this decreases the cost of borrowing money for municipalities, which allows them to charge less for the utilities they operate. The federal government, to offset the revenues it might otherwise derive from taxing the interest on municipal bonds, raises the general level of taxation on everybody. Swidler found this situation troubling. As an alternative to exempting the interest on municipal bonds from taxation, he argued that the federal government should directly subsidize public borrowers. Though Swidler did not fully think through this alternative (he implied that the federal government would use the revenues from taxing the interest on municipal bonds to pay for the proposed subsidies), the larger point he was trying to make was the need to apportion tax liabilities in a more equitable manner.

CHAPTER 14

Knoxville and Nashville, 1957–1961

There are few shocks in life as rude as losing your job when you do not have a new one, especially if you support a family. I was entitled to a federal pension of about $5,000, which was not insignificant in 1957, but was less than a third of the salary I had been paid. I had savings enough to carry the family for a couple of years, I was a pretty good lawyer, and I had some name recognition in the public utility field. On the other hand, I was fifty, and during the Eisenhower administration positions in government were closed to me. The obvious source of private employment was to work for a utility company, but I had no mind to switch sides in the public power fight. How to use my assets was a puzzle. As it turned out, opportunity laid unrecognized at easy reach.

It is more than a little difficult, even thirty years later, to acknowledge how little help I received from old friends. The circumstances of my departure from the TVA were not discreditable—on the contrary—but an involuntary termination marks a failure of sorts, and my case was notorious. I spoke with Lilienthal and Clapp in their New York offices, but they were no more than polite. I consulted an executive recruiter, John Corson, an officer of McKinsey & Company. His advice was to hitch my wagon to the Lilienthal star. Marguerite Owen, of the TVA Washington office, suggested that I live with my family on the civil service retirement allowance. Friends in private law practice in Washington could think of no way to be of assistance, though I had helped them in their careers. These rebuffs were painful and damaging. I have forgiven them all, if indeed they need forgiveness, because my experience in life has taught me that, for the most part, people are responsible for their own successes and disappointments.[1]

I was led into private law practice in Tennessee by a request for legal help from one of the TVA power distributors, the Joe Wheeler Electric Cooperative of North Alabama, a rural electric cooperative. Joe Wheeler was a desirable client. Unlike some of the cooperatives elsewhere in the country, which served only sparsely settled areas, most of the TVA cooperatives, with the TVA's help, took over private power facilities in built-up areas. Joe Wheeler was no exception. Its market area, bordering the Tennessee River on the south, was thriving. It profited from the industrial growth of Huntsville and Decatur, the two largest cities in the TVA's North Alabama operating area. Joe Wheeler itself was prosperous and growing rapidly. It served many thousands of customers, and its annual revenues ran in the millions. It was able and prepared to pay reasonable fees.[2]

The friction between Joe Wheeler and [the City of Decatur Electric System] was an example of an endemic problem. Joe Wheeler and most other electric coops were set up during the 1930s or 1940s, mostly in agricultural areas. With growth in population and income came industrialization and suburbanization. By the 1950s, many coops were serving areas that municipal systems (and, elsewhere, private utilities) now wished to serve, and frequently at lower rates. Cities sometimes strengthened their claims by "strip annexation" of the most attractive coop areas, which was maddening to the coops. Strip annexation consisted of annexing fingers of land along highways on which substantial development had occurred, without making city power (as well as water and other amenities) available to people in equally close-in areas whom it would be less profitable to serve. It was a form of cream skimming.

There was no perfect answer to such a conflict. Cities had a right to expand, and it would be wasteful to have two electric systems for one city. The coops also had equitable claims by virtue of taking the early risks of development.

The TVA was already concerned about the numerous territorial disputes that had emerged. Negotiating with the TVA power people (who had been my colleagues only months before), we worked out a formula for settlement. Joe Wheeler would sell the facilities in the disputed area to Decatur, but not at the offered price that I believed was close to the bare-bones book value of the facilities. Instead, Decatur would pay enough more to avoid increasing the cost of service for the remaining coop members. The parties finally agreed on the numbers, and the case was settled.

Several other cooperatives sought my help, on territorial disputes and other matters. I was working at home, and my wife did the necessary typing. I was reluctant to set up an office in Knoxville, which was not a central location for serving the TVA power distributors. A Nashville office would be preferable. Nashville was a larger city, with a more diversified business life and better opportunities for the growth of my practice. When the Tennessee Rural Electric Cooperative Association offered me the position of general counsel, provided that I moved to Nashville (where its own offices were located), I cast about for a Nashville affiliation. Through a friend, I made arrangements to share the offices (and the office costs) of an old Nashville firm, Denny & Leftwich.

My association with Denny & Leftwich was fortunate. The firm was small, with only five members, including the principals. Raymond Denny was a leader and former president of the Nashville Bar Association. He must have been in his early sixties when I joined. He was a shrewd and able trial lawyer and businessman, of country origins, without pretensions to legal scholarship. Louis Leftwich was the son of a former Nashville Bar Association president. He was an excellent lawyer who was also gentle and kind.

Private practice in Nashville was certainly different from TVA work. The atmosphere of the office was casual and the pace of work was relaxed. In the absence of deadlines, Denny and I would usually drop in at a nearby restaurant to drink coffee, meet other lawyers, and chat with local businessmen. The younger people in the office did not enjoy the same leisure. Typically, the older lawyers would break out a bottle of bourbon in late afternoon for a collegial drink before leaving for home and dinner.

Nashville proved to be the right spot for my practice. I soon acquired a number of new clients whom it would have been awkward to serve from Knoxville. They included the City of Memphis light, gas, and water divisions, the Alabama Electric Cooperative (a wholesale agency that served the Alabama distribution coops outside the TVA service area), the Tennessee Rural Electric Cooperative Association, the cities of Paducah and Princeton, Kentucky, and a major municipal bond house, Halsey & Stuart. I also became a consultant for the Public Works Committee of the U.S. House of Representatives.

The most interesting and lucrative matter I handled was the acquisition by Paducah, Kentucky, of the electric distribution system in the city, owned by the Kentucky Utilities Company (KU). I also assisted the efforts of Princeton, Kentucky, to acquire its distribution system from KU. This was a much smaller and simpler proceeding than the Paducah effort. The purpose of each was to enable the two cities to buy [cheaper] power from the TVA, that is, to become distributors of TVA power.

While I worked for the TVA, I had spoken on two or more occasions with the CEO of KU, Bob Watt, in an effort to negotiate the purchase of KU's west Kentucky properties, but with no success. He had pride in his company, an operation that served virtually the whole state outside the Louisville area. I understood his opposition, and the TVA, which was having trouble keeping up with the load growth of the existing market, left it to the Kentucky cities to make the fight for access to TVA power.

In 1958 or early 1959, a delegation from Paducah visited me in Nashville to solicit my help. I already knew something of the effort by Paducah to buy or condemn the [KU] distribution system. Indeed, it was a notorious case of a city's frustrated attempt to enter the electric distribution business. There were several reasons for the city's frustration, the most important being the skillful opposition of KU. The company was well managed, resilient, and served by able lawyers. It kept its political fences mended, partly by retaining leading law firms in many of the cities it served. It gave good service and was prominent in all activities to promote the attractions of its market area. In brief, hard fighting was supplemented by good works.

I suspect that another reason for the inability of Paducah to succeed in its acquisition effort was that the local politicians were not of one mind about the matter. This was certainly true at the time I took the case. The delegation that visited me in Nashville had assured me of the support of a united city council, but I found that Bob Cherry, the mayor, was strongly opposed. He did what he could to deter me, including strong expressions of personal hostility in a meeting that he requested. His hostility would have compelled me to withdraw if the other city officials had not assured me of their strong support. The mayor might have been able to stampede a city employee, but he could not stampede me. I was not willing to give him an easy victory, and I ignored him thereafter.

The success of the acquisition effort was no small victory for Paducah. The difference in rates between KU and TVA service was substantial. Paducah had been an important railroad town, though now its economic health was poor. Of course, shopkeepers and householders would pay less because of lower rates, but of almost equal importance was the city's desire to share the industrial growth that was transforming the TVA market. Large chemical plants were being built near the Kentucky Dam, using TVA power. Paducah benefited indirectly from this nearby development, and saw the benefits of placing itself on the same competitive footing as the existing TVA distributors.

The city was stymied because it was unsure of the best way to proceed. The franchise the city had granted KU many years before provided for municipal acquisition at the city's option. The city had filed a suit under the franchise, but the terms were general and the necessary procedures were complex. In litigation the company could push every technicality to its limits, as it had already done to the city's bafflement. They posed to me the question, which way should the city go? I pondered the question for a while and concluded that the better course was not to proceed under the franchise. The basic reason was not technical, nor even legal. The company had accepted the franchise. It was a contract under which the company's good faith was pledged. If there were ever a chance of a reasonable settlement, it would probably be in the franchise proceeding, to protect its reputation for keeping its word. Plus, a settlement had great advantages over a victory in litigation. In litigation, Paducah could acquire only the property within the city limits, which would force uneconomic breaking points on the distribution and transmission systems. In a settlement, reasonable boundaries could be agreed upon.[3]

The city fathers agreed and I plowed ahead. The first step in the franchise procedure required a hearing before a county court. I won at this stage, and ultimately that victory led to settlement negotiations and an acquisition contract. Perhaps one reason we were successful was that Congress, in August of 1959, had amended the TVA Act by restricting the TVA power-service area to that which existed on July 1, 1957, subject to a number of specific exceptions, including Paducah, Princeton, and four other Kentucky cities. In deciding to settle, KU undoubtedly took into account that Congress had sealed off the great bulk of the KU market from TVA incursions. At the time of the amendment's passage, I was serving as a paid consultant to the House Public Works Committee. I had great interest in the amendment because I helped prepare a basic draft of it in 1957. The final language [in 1959] was checked with me. I made sure that Paducah and the other cities were included among the exceptions. This did not involve a conflict of interest. My employment was public knowledge, the members of the committee were aware of my work for the Kentucky cities, and I had been employed specifically because of my background.

The acquisitions by the two Kentucky cities led me to consider again the basic economic justification for public power. The prices paid for the two distribution systems were more than twice their book values. These were not unreasonable prices in a settlement. In litigation, the courts might have set the prices even higher, and settlement avoided the delay, costs, and risks of carrying the litigation to a conclusion. Whether Paducah's political resolve would have been maintained during several more years of litigation was uncertain, considering Cherry's opposition. Moreover, KU was entitled to a substantial amount of severance costs for wrenching these two distribution systems out of an integrated utility system. The costs to consumers for carrying the extra acquisition costs (over book value) were not burdensome, considering that the introduction of TVA power would bring substantially lower rates. The costs above book value—described as "acquisition adjustment" costs—were amortized over a twenty-year period. Nevertheless, I had questions as to the fairness to KU and its investors, and whether the transfer was in the public interest.

It seems obvious that if nothing else had been involved, it would not have made any sense to incur the costs of severing a distribution facility from one integrated system and connecting it to another. The quality of service would probably not change. Employee costs were not likely to change significantly; Paducah and Princeton took over the KU working forces, except for a few people at the top. What changes in a typical takeover is not costs, but the transfer of costs. A municipality (not the TVA) can issue bonds on which the interest is exempt from federal income taxes; hence, its bonds are more attractive to investors than corporate bonds, though they carry a lower yield. The fact that municipal bondholders make no contribution to the costs of the federal government does not mean that these costs have been saved; instead, other taxpayers must pick up more of the total. These are called "transfer costs." The loss to the U.S. treasury as a result of funding the public systems with tax-exempt bonds is not small. Vast amounts of such debt, many billions of dollars, have been issued by states, cities, and power districts. Of course, few individual cities issue enough tax-exempt bonds to make a perceptible difference in the national debt or on the burden of other taxpayers. Cumulatively, though, tax-exempt electric revenue bonds are a significant contributor to national deficits. Similarly, a municipal system typically makes no payment of state taxes, and only negotiated amounts in lieu of taxes to the city. Here, too, there is no savings in the cost of government, but only a shift from one group of taxpayers to another.

Why does the income tax exemption on revenue bond interest endure, in the face of the manifest waste and unfairness? A strange group of bedfellows has a financial stake in the question. The investment houses that issue the bonds are among the largest in the country, and bond sales account for a substantial part of their profits. Wealthy individuals who buy the bonds cherish this income tax exemption, which has little or no value to people of small means, most of whom do not pay federal income tax or pay at low rates. The corporate trustees who manage the financial affairs of the wealthy have a similar interest. And, of course, the states, counties, cities, and innumerable public districts with bond issuing authority save much of the cost of money that they would otherwise incur in their subsidies to large and profitable businesses, which are the biggest gainers.

For the electric systems, the revenue bond tax exemption means more than savings in the cost of borrowed capital. In effect, they are in constant competition with investor-financed systems, or so they think. The tax exemption provides them with an advantage. It protects them from being acquired by the private systems. Conversely, it also provides the basis for the claim of rate savings when cities and other public agencies seek to take over the properties of the investor-owned systems.

Not merely for "yardstick" reasons, but for a broader reason of public policy, I have been opposed to this tax exemption feature for many years. It seems to me, and to many others, that the federal income tax code lacks rationality and fairness, that it lends itself to loopholes and discriminatory exemptions, and that some people of large means pay no taxes while others of lesser means pay too much. Elimination of tax-free bonds is the key to tax reform. The case is strong because tax-free bonds are an inefficient way to reduce the cost of borrowing by state and local agencies. Far better would be an outright subsidy to public borrowers to offset the difference in the cost of money between taxable and tax-exempt offerings. In this way, the cost to the treasury would be definite and visible. It would also be less likely to foster a contagion of tax preferences.

The Phillips Letter

While I was living in Nashville, I learned something about the political tactics of one of the oil companies that had extensive transactions regulated by the Federal Power Commission. In my new affluence, I had bought some shares of the Phillips Petroleum Company. Phillips was widely known to be a political hard-liner, but I was buying its shares as a financial investment, not as a political endorsement. When the 1960 presidential campaign was heating up, Phillips sent a letter to stockholders warning of the dangers to the oil industry if Democrats won; it strongly urged stockholders to vote and campaign for Republicans. Possibly the letter came later, after the party conventions, and referred by name to Kennedy and Nixon. At any rate, as I recall, the letter specifically discussed the IRS's "depletion allowance" for oil and gas operators, and the danger that the rascally Democrats, if they came to power, would deprive the oil and gas industry of this tax plum.[4]

I was incensed when I received the letter. I thought it was wrong for the company to use stockholder money to tell stockholders how to vote. I was exercised enough to send a letter, along this line, to the chairman of the board of the company. If I received a response, it was not remarkable. A few weeks later a Nashville colleague asked me what heinous injury I might have inflicted on Phillips Petroleum; it seems that Phillips had approached him and some other Nashville lawyers, inquiring whether I consorted with known enemies of the oil and gas industry. I told him about my letter. He grinned.

The incident was not funny, as it says something about corporate influence on political destiny. I was not a person known at that time to the officers of the company, I held no official position, and I could rant daily on the street corner about the inequity of the depletion allowance without the slightest effect on anyone who passed. But to the government relations staff of Phillips, any sign of disagreement was too much.

The incident had an aftermath. Some months later, Kennedy was elected. Before he took office, he announced his intention to nominate me as chairman of the Federal Power Commission. Phillips had identified me as someone potentially threatening to its interests, and shortly after Kennedy was sworn in, while I was still in Nashville, Phillips officials visited the White House to urge the president to change his mind about appointing me. The next day, I received a call from Ralph Dungan, the president's chief of staff, to ask me about the complaint. He said Kennedy was concerned. Apparently, the Phillips representatives based their opposition to me on the claim that in the letter I had attacked the depletion allowance, and that the letter served as a symbol of my general opposition to the company and the oil industry. They did not produce a copy of the letter. Actually, the letter did not mention the depletion allowance, and, for lack of occasion, I had never expressed my view to anyone on that issue. I explained to Dungan that the letter was solely a plea that elections should not be tilted against Democrats by corporate appeals on behalf of Republicans, paid for by stockholders. I sent him a copy of the letter, and that was the end of the matter. Ralph reported to me that the president enjoyed reading my letter and said it confirmed to him that I was the right man for the FPC job. Before I took office, I sold the Phillips stock and whatever else in my pitiful little portfolio might be construed to create a conflict of interest.

Joseph Swidler as a young lawyer in the 1930s. Courtesy of Ann Swidler.

Joseph Swidler conferring with David Lilienthal in the 1930s. Courtesy of Ann Swidler.

Joseph Swidler with Senator Warren Magnuson, center, and Senator Gordon Allott in 1964. Courtesy of Ann Swidler.

Joseph Swidler and Vice-President Lyndon Johnson in the early 1960s. Courtesy of Ann Swidler.

Joseph Swidler inspecting Consolidated Edison, New York, facilities in the early 1970s. Courtesy of Ann Swidler.

Joseph Swidler and his wife Gretchen in the 1980s. Photograph by Carol Harrison. Courtesy of Ann Swidler.

PART VI

Chairman of the Federal Power Commission, 1961–1965

In January of 1966, the *New York Times* provided a summary of Swidler's performance as FPC chairman: "Mr. Swidler . . . transformed the Power Commission from one of the most inept agencies in Government to one of the best . . . [and] finally achieved the impossible: He left with the praises of both consumer groups and industry ringing in his ears." This assessment was not an exaggeration. Swidler reformed an agency that had a justifiably dismal reputation, and he did so in a manner that pleased utility executives as well as the average citizen. It was perhaps his greatest achievement in public life.[1]

Swidler's description of his transformation of the FPC tells us much about the daily operations of administrative governance. It also suggests how important the vision and leadership of a single individual can be. It is not, however, a story of drama and derring-do, but an account of the way seemingly modest, though significant, changes altered the direction and enhanced the efficiency of a key federal agency. As a result of these changes, the FPC came to define the limits and potential of modern regulation.

As FPC chairman, Swidler did not shy away from the rough-and-tumble of politics. He attempted to cultivate political allies and appease likely foes even prior to his confirmation—though telling the difference between the two groups was sometimes difficult. Complicating matters, Swidler was confronted by a sitting FPC chairman who did not recognize Kennedy's authority to remove him from office. That was just the beginning of Swidler's political education. Throughout his chairmanship he had to distinguish between what was desirable and what was possible.

This meant acknowledging the allegiances of his fellow commissioners, recognizing the power Congress had over appropriations, and understanding how far to push certain reforms.

Political realities did not prevent Swidler from implementing the changes he deemed necessary. Basic considerations—for example, the arrangement of the commissioners' chairs during meetings—were followed by more substantive modifications. Swidler significantly increased the size of the FPC's staff; he encouraged the commission to initiate its own rate complaints; he established an economics division within the commission; he streamlined procedures so that the commission could handle its case load more effectively; and he prohibited the commission and industry representatives from discussing pending cases in private. Taken together, these changes permitted the FPC to assume a far more active and effective regulatory role.

Swidler also introduced changes that specifically targeted the commission's regulation of natural gas and electricity. During the 1950s, some members of Congress had pushed for deregulation of the natural gas industry. A sympathetic FPC aided that cause by allowing countless rate cases to languish on its docket. This was one of the reasons the FPC's reputation had become so tarnished. Swidler moved swiftly to remedy this situation. He introduced an expedited decision-making process and enforced an area-based rate schedule that allowed the commission to establish rates for whole groups of producers, instead of determining them on a lengthy case-by-case basis. Swidler's tenure would have likely been praised if these had been the only changes he implemented.

The innovations Swidler made in the FPC's regulation of electricity were equally significant. He mandated new safety requirements for hydroelectric dams and insisted that federally owned dams develop on-site recreational facilities. Furthermore, at Swidler's urging, the FPC settled tricky jurisdictional issues and published an important analysis of national power resources. His influence also led to the creation of an electric power research institute. Such an institute seemed all the more necessary after the Northeast blackout of 1965, the largest power outage up until that time. Under orders from President Johnson himself, Swidler had to ascertain the blackout's causes and recommend steps to prevent a similar occurrence from happening again. The actions he took during and after that emergency illustrated his ability to solve problems in a calm and decisive manner.

Overseas travel comprised another important aspect of Swidler's FPC duties. In 1962, he had the opportunity to participate in an official mission to the Soviet Union. His observations not only highlight the ideological tensions dominating U.S.-U.S.S.R. relations during that era, but they also outline the strides that the U.S.S.R. had made in certain areas of power production. Swidler visited London as well. The investment bank of Model, Roland, &

Company paid for that trip. Swidler's purpose in going was to give a speech to a large group of European investors about the advantages of purchasing stocks in private American utilities. Swidler believed that additional investments—by increasing the supply of capital—would help utility companies implement newer, more efficient technology; this technology, he felt, would translate into lower rates for the consumer. Though Swidler's reasoning was logical, the appropriateness of his speech, in which an administrator endorsed investment in certain companies, seems questionable. It is the only point in an otherwise self-conscious and often self-critical memoir that Swidler appears to have been unaware of how his actions might have been construed, if only retrospectively, in an unfavorable light.

Swidler's own evaluation of his FPC service does not coalesce into a concise regulatory philosophy. He would have certainly agreed with *Business Week*'s description of his leadership style: "Swidler likes results more than strife over doctrine. He leans to negotiation and compromise. He loves novel solutions to tough problems—the kind with something for the consumer, and something for the company." His actual philosophy was composed of similarly diverse elements. Rather than directly focus on lowering rates, Swidler promoted strategies that would allow utilities to reduce costs; he hoped—sometimes naively, but often accurately—that utilities would pass these savings on to the consumer. For this reason, Swidler advocated measures that would rationalize the production of gas and electric power. A related objective was integrating the nation's electric power systems. Integration would lead to economies of scale and enhanced productivity, which, according to Swidler, would result in lower electric bills. In sum, Swidler's views did not represent a simplistic company versus consumer, or even public versus private power, philosophy. Swidler admittedly erred on the side of the consumer, but in doing so he preferred to offer carrots, not sticks, to utility companies. It was this concern for fairness and equity that emerged as the strongest component of Swidler's regulatory views. It was also the reason he left the FPC with so much praise "ringing in his ears."[2]

CHAPTER 15

Back to Washington, D.C., 1961

One day in late November or early December of 1960, when I was in Lexington, Kentucky, working on the final arrangements for the KU settlements with Paducah and Princeton, the White House switchboard tracked me down for a call from two members of President-Elect Kennedy's transition team, Ralph Dungan and Adam Yarmolinsky. They came quickly to the point: Kennedy offered me the chairmanship of either the TVA or the Federal Power Commission (FPC). Would I accept, and which would I prefer? I was very surprised. I had taken no part in the Kennedy campaign, or any other, except to vote; I had never participated in party politics, nor had I any intimation that I was under consideration for an appointment. In fact, I had assumed that the awkward circumstances of my leaving the TVA would preclude further service in the federal government. After a short pause, I said that more than twenty years of working for the TVA were enough, but that I might consider the FPC chairmanship.[1]

There were lots of reasons to decline the offer. I was now well established in a busy practice. I shared offices with people I liked and with whom I had easy business and social relations. Gretchen and I were beginning to become involved in Nashville civic work. We had a beautiful and spacious home, better than we were likely to be able to afford in Washington. I was earning more than $50,000 a year, high earnings for lawyers in Nashville in 1960. There was only one argument on the other side, but in the end it outweighed all the others—the opportunity to do a good job in a high and challenging position for which I believed I was qualified. I had worked hard in the shadows of other men, preparing letters for their signature, submitting papers and ideas for their consideration, and providing advice that they could accept or disregard. Now I had the chance to take a lead role in the Washington limelight.

In due course, I phoned the two Kennedy men. First, I expressed a preference for the chairmanship of either the Federal Trade Commission (FTC) or Federal Communications Commission (FCC). Most of my government service had involved energy policies, but I had broader interests and would have liked more diversity in my career. I was told that both were spoken for—the FTC job to the chief aide of Senator Kefauver, Paul Rand Dixon, and the FCC spot to a Chicago lawyer, Newton Minow, a protégé of former Governor Adlai Stevenson of Illinois. I was disappointed because both jobs were in areas of more general scope and influence (and prominence, as well, to be candid) than the energy position. I had fantasies about the reforms I might institute to enable those agencies to serve the purposes of their creation better. But my ideas for improving the FPC's performance were better grounded and more concrete, my acquaintance with industry leaders infinitely greater, and my chances for a successful term of office probably more promising. At any rate, I accepted the FPC job, and a few days later Kennedy announced his intention to nominate me.[2]

Truth to tell, the FPC was not then a glamorous agency, and a new chairman was in for a hard time, whoever he might be. In a study made for Kennedy, James Landis, a former dean of Harvard Law School and a noted expert on administrative law, called the FPC the poorest managed of the six major Washington regulatory agencies. It exercised authority over the interstate transactions of two basic energy industries—electricity and natural gas. On the electric side, it had become a cipher, while its inefficiency on the gas side had created a backlog of cases.[3]

In preparation for my new job, I spent a great deal of time in Nashville reading all the natural gas opinions of the Supreme Court and the important appellate cases. I prepared summaries of the cases and gathered the summaries into a notebook. When Larry O'Connor joined the FPC and discovered that such a volume existed, he borrowed it and found it useful. He passed it on to other commissioners as they came on board. I was already reasonably familiar with the electric power case material.[4]

Among the most enthusiastic of all my friends in Nashville was Jimmy Riddle. He was, or had been, in the coal business and had made a number of sales to the TVA. Pleasant and deferential to the point of excess, he was a persuasive salesman. It turned out that he had made a loan to the TVA's coal buyer, who resigned when the fact came out. When I moved to Nashville, Riddle pressed me to be his friend, for reasons I could not understand. On two or three occasions, over as many years, we dined together and he had always insisted on picking up the check. It is hard to turn down offers of friendship, even from someone with whom you have little in common, and about whose business ethics you may have some qualms, especially if he is lavish in his praise, assiduous in his pursuit, and does not display any ulterior motives. I did not see at the time any way I could possibly be of help to him. I was at first merely polite and, later, sufficiently receptive to meet with him occasionally. When my prospective appointment was announced, a great change took place. He became a fountain of unsolicited advice on the appointment process. He visited me several times at home to counsel me on how to handle myself in a variety of difficult circumstances that he envisaged, mostly related to Senate approval. It was not until he turned up later in Washington that I recognized a possible economic basis for the intensity of his interest. But more about that later.

I managed while still in Nashville to get estranged from Drew Pearson, the country's most widely syndicated columnist. Pearson phoned me from Washington to ask me what I proposed to tell the Senate Public Works Committee, to which my nomination was referred, as to the circumstances of my leaving the TVA and other matters. I answered, with my best effort at diplomacy, which was not a native skill, that I wanted to be responsive, but that I owed it to the members of the committee to tell my story first to them. He was not to be put off with fair words and spoke in a peremptory and threatening way in an effort to force me to make disclosures. I tried, without success, to placate him without yielding ground. It was a lesson in how some journalists recruit their informants and gather their scoops. I paid a heavy price for incurring Pearson's displeasure. In four-and-one-half years at the FPC, I never received favorable mention in his column; he printed many critical stories. Columnists evidently cultivate long memories. Readers relying on his column would never suspect that I was doing a decent job and was receiving praise from other sources.[5]

At the same time that the president announced my intended appointment, he also announced his intention to appoint Howard Morgan of Oregon as a member of the FPC. It turned out that Morgan was a bountiful source of inside information and rumors for Drew Pearson. I began receiving phone calls from Morgan in Oregon, forecasting great troubles for me in the confirmation hearings, and urging me to tell all to Pearson, which would somehow protect me from the forces of evil gathering around me. The forecasts of doom did not bother me. I had a good story to tell, my conscience was clear, and I had long experience in testifying before congressional committees. The real danger was in affronting the Public Works Committee by blurting out information in advance of the hearings, especially to Pearson, whose standing with most members of Congress was abysmal. Morgan's calls, frequent and prolonged, were a strain because I was compelled to reject his advice, yet I needed to remain on good terms with a future FPC colleague.[6]

A serious blow was yet to fall. Sometime in the early spring, I was told there was a little hitch. The present FPC chairman, Jerome Kuykendall, a Republican, denied the right of the president to remove him as chairman. He had evidently cleared his act of defiance with the Republican senators. Here was a confrontation to the taste of the press, and they made the most of it. It had never occurred to me that the Kennedy people had not determined the existence of a vacancy before they had offered the job to me and announced the choice to the world. I quickly researched the question and found that there was indeed uncertainty in the law. I concluded, however, that Kennedy had a good shot at winning the argument if he held his ground. I was told that the White House was working on the problem and to sit tight.[7]

Sitting tight was difficult. I decided to cast the die and move to Washington so that I could help in clearing up the appointment mess. I had not sought it, and could have been happy without it, but at this stage my pride and dignity were involved and I saw no alternative to seeking forceful action from the White House in support of the president's nomination. My physical presence would act as a prod to force the president and his aides to address this unwelcome problem. I had some familiarity with the legalities and procedures for the removal of presidential appointees, as a result of my participation in the removal of A. E. Morgan [from the TVA] by President Roosevelt in 1938. I thought my experience might prove useful. We moved to Washington—wife, children, and ailing sheepdog—to a rented house on Garfield Street, near Wisconsin Avenue, in the spring of 1961.

During the mid-1940s, Roosevelt had ordered the army to take over the offices and other facilities of Montgomery Ward, then the nation's largest retailer after Sears Roebuck. The head of the company, a stubborn old man named Sewell Avery, had denied the legality of the action and refused to leave his office. Roosevelt responded by having soldiers carry him out. The newspaper pictures of the occasion were unforgettable—a little old man, silver hair neatly parted, sitting undaunted in his desk chair as he was carried to the street by burly soldiers. The scene was both affecting and comic, and in 1961 nearly everyone must have remembered it. This was the image the Kennedy administration was anxious to avoid. Kuykendall had even less desire to play Sewell Avery to Kennedy's Roosevelt, but he did not let on.[8]

Fortunately, I had a friend in the White House named Lee White, whom I had hired for the TVA legal staff after he graduated from the University of Nebraska Law School. White—sensible, amiable, good company, and a talented storyteller—was a White House favorite who did me many favors. I lunched with him often at the White House mess, and, as a result, came to know the president's intimate circle better than most department or agency heads. White was also helpful with the Kuykendall imbroglio.[9]

I had one or two talks with Ted Sorensen in the pre-confirmation period. I remember only one subject of discussion, a proposed reorganization plan for the FPC. Under an administrative reform initiated by Eisenhower, the president had been granted temporary power, which had been carried over into Kennedy's term, to reorganize the administrative agencies, subject to congressional approval. The FPC reorganization plan had been fully drafted on the basis of recommendations by Landis. In my view, the plan was not a good one. I asked [Sorenson] to keep the draft open for revision for a few days so that I could study it in detail and provide him with a written report of my views. He seemed nettled by the suggestion and declined it on the basis that he had agreed with the president on a timetable for producing reorganization plans. If the FPC plan were delayed, he would be thrown off schedule. I was surprised that he cared so little about the views of the chairman-designate on whether the plan would help or hurt the agency. He did not ask me to support his plan, apparently more concerned with its submittal than with its practicality or its acceptability to Congress. I did not criticize it, but I felt forced later to seek and obtain more workable reorganization solutions.[10]

I was instructed early on by the White House to visit key senators and make myself available for any questions they might wish to ask. I did so, but I have a poor recollection of most of these sessions. The one that stands out in memory was the interview with Senator Robert Kerr. He was the Kerr of Kerr-McGee Oil Company, one of the largest natural gas producers in the country. He had entered politics at the top—after amassing a large fortune—and was then the chairman of the Senate Rivers Subcommittee. A huge man in height and proportionate girth, he was an effective speaker and able debater on the Senate floor. He was one of the half-dozen most influential members of the Senate. He had three goals: (1) to secure the maximum appropriations for public works development in Oklahoma, including a project for bringing a navigable channel from the Mississippi River to points in Oklahoma; (2) to protect the interests of the petroleum industry in general, and Kerr-McGee in particular; and (3) to harass and discomfit the electric power industry. The mixture of populism with financial and industrial aggrandizement fit this oversized and seemingly homespun man of large business affairs.[11]

An accomplished player of the political game, Kerr used great care in espousing the causes of the oil and gas industry. He assumed a low profile in this phase of his work, or as low as was possible for a man of his stature. Nevertheless, he did not avoid the spotlight. A cynic might think that his anti-utility position was only a foil for his activities on the petroleum side, but I did not doubt his good faith, and he seemed to put his heart and sardonic wit into his utility baiting, which I confess I much enjoyed when I worked for the TVA. Several times I watched as he cross-examined a utility witness, typically representing the Edison Electric Institute. His pleasure in his ability to confound and demolish the witness was evident. At any rate, he was implacable, and I had no desire to antagonize him.[12]

The interview with Kerr went off much better than I had expected. The key question was my view on natural gas prices. He danced around this topic to give me a chance to blurt out a clue to my thinking, but I could tell him with candor (when he finally asked me) that I was far from having the answer on either substance or procedure on gas pricing. I assured him, in response to another question, which doubtless derived from the Phillips Petroleum letter, that I had never taken a position on the depletion allowance. Nor did I believe that this tax question could ever be an issue before the FPC. During the interview, Kerr treated me respectfully, said nothing untoward, and neither then nor later did he seek to influence me, even by having an assistant call me for information—which is more than could be said of his Oklahoma colleague, Senator Monroney.[13]

My meeting with Senator Jacob Javits was also memorable. We talked for only a minute or two in one of the cloakrooms off the Senate chamber. Javits was in a hurry to return to the Senate floor. He said he knew something about my qualifications, and was satisfied. He had only one question—was I Jewish? I assured him that I was indeed a Jew.[14]

The confirmation hearings for Howard Morgan and me were held jointly, beginning on April 11, 1961, before the Senate Committee on Interstate and Foreign Commerce. On the first day, Senators Kefauver and Gore, both highly respected in the Senate, testified for me. Kefauver's statement was particularly strong. Senator Pastore, serving in the absence of chairman Warren Magnuson, announced that Senator Lister Hill, also greatly respected and much-loved, endorsed my nomination.[15]

The questions from the committee members enabled me to give my views on many problems that would likely come before the FPC. The most obvious question was whether I could be fair to the privately owned utilities, considering my public power background. My answers were emphatic. I responded that I was not taking the job as a public power advocate, but would be guided only by the provisions of the Federal Power Act and the Natural Gas Act. I thought the 1935 legislation had largely cured the electric utility industry of the faults that had brought it into disfavor with the public in the 1920s and 1930s. I had no animosity against the utilities; on the contrary, it would be my aim to strengthen them. The FPC should take the initiative to reduce rates that were excessive—not merely vote on requests for rate increases. The independent agencies should be regarded as arms of Congress, but, of course, should work with the executive branch when their independence was not compromised. It would be proper for a member of Congress to make inquiry as to the status of a proceeding, and I would respond to such an inquiry. Overtures from the White House with respect to cases in litigation were improper and would be repulsed.[16]

On most of my answers, I had no reservations or qualms. Others I was able to answer positively because of the way the questions were framed. Probing, though, would have revealed some uneasiness, particularly questions that might have dealt with the relations between the administrative agencies and members of Congress. A member could not be denied the right to know the status of a case, but only a political innocent would believe that the purpose of a call to the chairman was to secure routine information. The point of such a call was to impress upon the chairman the representative's interest, and to satisfy a hope or expectation that this would lead to quicker or better treatment. This problem was defused to a great degree during my tenure by the adoption of a rule by the FPC that required all such inquiries and responses to be made a part of the record of the case.

I have mentioned Jimmy Riddle, my Nashville friend, who had developed a passionate interest in my career. On the morning of my Senate confirmation hearing, a funny thing happened on the way to the Capitol. My wife and I were staying at the Washington Hotel. As Gretchen and I stepped out of an elevator into the hotel lobby on our way to find a taxi, there was Jimmy, accompanied by a friend. He rushed up to us, announced that a limousine was waiting, and, over our weak protestations, hustled us into it. We were introduced to his companion, an executive of the Tennessee Gas Transmission Company, which operated the country's largest gas transmission system. This was my first clue concerning his interest in me, but I had no time to think about it. At the conclusion of the hearing, he provided the same transportation service. I told him that Gretchen and I had other plans for the evening. I suppose he was demonstrating to Tennessee Gas ("TGT"—but not to be confused with Texas Gas Transmission Company) the depth of our friendship and his ability to influence me. He was driving me to the conclusion that I had had enough of Jimmy Riddle; I still thought of him as a Nashville friend, though, and did not want to hurt his feelings. The issue came to a head soon after I was sworn in. Riddle invited me to dinner and said he wanted to bring a TGT friend. I said, "No friend." We ate alone, mostly in silence, but during dinner he pressed me to allow the friend to join us. I was firm. It was not a cheerful meal, and I never saw or heard from Jimmy again. He had picked the wrong man.

Eventually, Congress approved my and Morgan's nominations, and the time approached for swearing-in. At my suggestion, the president had asked for an opinion from the attorney general on his right to designate a chairman. The request was referred to the Justice Department's Office of Counsel, which prepares responses to such a request. The man who prepared the draft, Harold Reis, an able lawyer, gave a prompt reply. The attorney general formally advised the president that the power of appointing a chairman was his. There was newspaper speculation as to whether Kuykendall would be carried out of his office, chair and all. (Actually, the plan was not to carry him out of the building; instead, his things would be moved from the chairman's office to another space—he was to remain a commissioner after losing the chairmanship.) The president did not want a confrontation over a matter that might make him a subject for ridicule and complicate his relationships with the Republican senators, but retreat was impossible. Behind the scenes, the White House negotiated with Republicans, upon whose instructions Kuykendall was acting, and a face-saving compromise was reached on the eve of my induction. Kuykendall was to stay on as chairman until September 1, a little more than two months, and would then yield his chairmanship. This arrangement, which was made public, suited me perfectly. It would take a couple of months for me to get my bearings. In the meantime, the other commissioners and the staff would treat me as the chairman to the extent that I wished. I had no desire to embarrass Kuykendall. On the contrary, in the course of the negotiations, I had come to like him.[17]

I was sworn in on June 28, 1961, by Judge Charles Fahy of the Circuit Court of Appeals for the District of Columbia, under whom I had worked during my brief stint in 1933 in the solicitor's office of the Department of the Interior. His participation added to the pleasure of the occasion for me and my family.

CHAPTER 16

Turning the FPC Around

The first commission meeting I attended provided dramatic notice of the critical need for procedural reform. It was held in a small room off the large commission hearing room. Kuykendall presided, since he was still chairman. I attended, but for lack of familiarity with the agenda, chose not to vote, although I spoke up at the end. Only commissioners were present, except that the FPC's secretary, Joseph Gutride, was on call and came in from time to time to answer a question or supply a document. Kuykendall introduced each item by caption, providing only a cursory explanation of the nature of the item, and without mention of any significant questions that might be involved. There was little discussion, and all agenda items were quickly approved. I was appalled. Here was the perfect setting for skullduggery, and it certainly lent itself to that suspicion. I said politely, but firmly, that the meeting format must be changed. I described as a model the TVA board meetings, attended not only by the directors, but also by the general manager, the general counsel (who was also the secretary), an assistant secretary to prepare drafts of the minutes, and, most important, the heads of all the departments and offices.

Procedural Changes

In the TVA model, staff people were invited to express their concerns, and the discussions were often lively. The staff not only received guidance at these meetings, but also became aware of everything important in the whole organization that was going on or planned. As a result, they were better able to coordinate their programs. They served as conduits for both policy implementation and for bringing employee and public viewpoints to the board. The board, in arriving at decisions, was able to use the tremendous reservoir of knowledge and wisdom of the key staff people around the table. If the FPC did not use its staff resources in deciding policies and cases, what was the point of having a staff? The other FPC commissioners expressed a willingness to adopt the proposed arrangement, and that is the way the commission meetings [were] carried on from that time.

After reaching agreement with the other commissioners on the format for meetings, I addressed another problem. I thought the meeting arrangements should be functional. The FPC, in large meetings, sat at a rectangular table at the same elevation as the rest of the commodious meeting room, so that not all commissioners could see each other. I proceeded with rearrangements to give the room something of a courtroom appearance, with the commissioners on a low platform and their seats arranged in an arc so that they could face each other. In turn, everyone in the room could see all the commissioners. These were not insignificant actions. They added to the dignity of the commission and facilitated communication. They paved the way for the kind of vigorous interaction upon which good administration depends.[1]

More important were the new assignments for the staff. In the past, the staff rarely took an active role in the trial stage of proceedings. FPC lawyers and staff attended the hearings, but staff did not testify, and the lawyers limited themselves to occasional cross-examination, usually based on advice from engineers or accountants. In effect, it was left entirely to the parties, the buyers and sellers, or the applicants and intervenors, to create the record upon which the FPC's decisions would be based, much like a court. A commission, though, is not a court. A regulatory agency is a guardian of the public interest, and not merely an adjudicator of disputes. Thus, a judge in a law court has no staff, except for law clerks, and is presumed to be entirely neutral in disposing of controversies. The FPC had more than a thousand employees, including hundreds of technical experts. A judge may honorably reach a wrong answer, if it is the one supported by the record. In a criminal case, for example, the defendant may be acquitted, even if the judge or jury believes him to be guilty, if the record does not prove guilt beyond a reasonable doubt, or if an important procedural error was committed by the prosecution. But a regulatory agency is provided with its own expert and investigative resources to insure that it reaches a correct conclusion, consistent with the facts and the law, and governed by the standard of the public interest.

The FPC was given authority to initiate cases on its own motion—to reduce excessive electric or gas rates, for example. Here, the FPC itself would be the complaining party. There might not be any testimony to refute that of the respondent company unless the commission took the initiative. Of course, public agencies might intervene, but this was not assured. The utility field is too arcane, and expert witnesses too expensive, to assume that in all cases intervenors will adequately address the testimony of a utility company. In some cases, the FPC had mounted a case of sorts, but only erratically and usually only by cross-examination. The commission had the burden of trying to find the right answer from two usually contradictory presentations, without benefit of testimony from staff on a reasonable and sustainable basis.

I gave instructions that it was the responsibility of the FPC staff to put on a public interest case in every proceeding that went to the hearing stage. No longer were they to be hand wringers or complacent observers. They were to mix in the fray and attempt to insure that the record supported an outcome in the public interest. The staff accepted this new assignment with satisfaction and zeal. Active trial experience honed the talents of the staff, added pleasure and excitement to their work, and increased the value and marketability of their services. From that time, the staff case and recommendation were principal resources for FPC decisions, not because the commission gave staff a preferred status—it did not—but because staff positions frequently proved to be less extreme and more sensible than those of the other parties. This arrangement has become almost universal among public utility regulatory agencies, state and federal.

When I became chairman, a procedure was already in place whereby the conclusions of hearing examiners (or administrative law judges—ALJs) were forwarded to the FPC as recommended decisions. The FPC made the ultimate decision, on the basis of claims of error by the parties, called exceptions, which were supported by briefs. FPC decisions, in turn, were subject to a second review. Final decisions were themselves subject to court review. While the new commissioners made no change in this scheme, they took the responsibility for careful analysis of the record and arguments more seriously. Frequently, on rehearing a case, the commission made extensive changes to its initial opinion. This was not strange because the first decision might depart significantly from the examiner's recommendations, and only on rehearing could the parties address this new material.

The record in important cases might occupy thousands of pages. Exhibits could well number in the hundreds, and briefs could make a stack at least a foot high. It seemed clear to me that the FPC needed help with its workload. Additional individuals were recruited, for the most part from other FPC offices. On the basis of their analytical gifts and writing ability, they were called the "opinion writing" staff. The chief of this group designated a member to be responsible for helping the commission handle cases. That individual was to analyze the record, prepare summaries for the commissioners, identify the principal questions, suggest and comment upon the range of answers, and, after meeting with the commissioners, prepare a draft opinion. The opinion writing staff proved to be an important and successful innovation. It consisted of only a few people, perhaps six, and they were key to the FPC's ability to handle its workload.

Many of our cases were complex, and some were of great importance. It seemed to me the FPC could better understand and decide these cases if it had the benefit of oral arguments; therefore, we heard several cases each year in this manner. I found the arguments helpful, and they were popular with counsel and the litigants, who were able to appraise their case, the effectiveness of their counsel, and the work of the commission.

The hydroelectric licensing cases were a special problem. The statute did not require a hearing in non-contested hydro cases, and usually none had been held. The provisions of the licenses were, in effect, negotiated between the applicant and staff, perhaps with behind-the-scenes intervention by commissioners. The results were submitted for pro forma approval by the commission. If there were strong protests or interventions, a hearing would be held, but the staff usually preferred a quiet negotiating process. This arrangement did not suit me or the other commissioners at all. We decided there must be a public hearing in every hydro case, with staff participation on the record.

I found when I came to the FPC that even though each of the opinions was signed by a commissioner, who took nominal responsibility for its preparation, it was usually prepared by the ALJ who heard the case, or by a commissioner working with the ALJ. This seemed to make a mockery of the review by the commission of the ALJ's decision. As a practical matter, the supervising commissioner had the opportunity to skew the commission action in favor of sustaining the ALJ's recommendation. Moreover, the assignment of responsibility to a single commissioner to prepare a draft opinion was made before the commission had decided the case, and usually without instructions from the commission as a whole on the disposition of the appeal. The commissioner responsible for the draft opinion had an advantage over the other commissioners because he had, or could claim, a more intimate knowledge of the case than his peers. He tended to be defensive in discussions of the product he presented (the day of the female commissioner had not yet arrived), and there was more deference to the opinion-writing commissioner than was healthy for the soundness or fairness of the result, let alone the scholarship and the literary quality of the opinion.

The new commissioners put an end to this arrangement. Assignments to a commissioner to prepare or supervise the preparation of opinions were to be made only after the whole commission had decided the issues in the case. The judicial independence of the ALJ was not to be compromised by requiring or permitting him to assist in the preparation of the commission's opinion. Nor were commissioners permitted to influence the ALJ's recommendation. The ALJs were accorded, as much as possible, the independence of federal judges. As a result of these reforms, debate among the commissioners was vigorous on every phase of an opinion, and drafts prepared under the supervision of any commissioner, including the chairman, received intense and unsparing scrutiny from each commissioner able and willing to take the time to study the drafts.

Before my time, commissioners had only their secretaries to help them. At their request, I authorized each commissioner to employ an assistant of his choosing. As chairman, I had two assistants, since the administrative burden fell upon me. All these assistants were of high caliber, and proved to be indispensable. Later, the other commissioners requested a second assistant. I thought this was too much of a good thing. Assistants were diligent and fertile, and the more people involved in reaching a decision, the more the emphasis on minutiae and the more difficult and extended the review process. Assistants also created administrative distractions for the commissioners. I declined the request for a second assistant.

A number of other procedural changes were made. To insure fairness, staff members who participated in a trial or preparation for a trial were excluded from commission meetings when the case was discussed. The trial staff could meet with representatives of the parties and discuss the merits of a case, or possible settlements, but no one in the decisional chain—ALJ's, FPC witnesses, the trial staff, the opinion writers, commissioners and their personal staffs—could talk privately with representatives of parties in contested cases without violating the commission's ex parte rules. We tried to assure that the judicial side of the FPC's work was totally protected from improper contacts with any of the parties.[2]

The ex parte rule did not prevent discussions among people in the decisional chain or representatives of the various parties, so long as they did not deal with pending cases. Since the major gas companies usually had several or many proceedings pending before the commission at any given time, a rule prohibiting all contacts with a company's representatives would have cut off virtually all communication with the industry. I met frequently in my office with industry representatives and business executives themselves (not their lobbyists or "government relations" vice-presidents, for whom I had little patience), listening to their views on problems other than those in litigation. My door was open to all segments of both industries, and to consumer representatives, without favoritism. These meetings provided interesting perspectives and information on the structures of, and internal relationships within, the two industries. I was strict in adhering to the ex parte rules. I had a copy framed and hung conspicuously in my office. If a visitor began to veer toward impropriety, I would point to the rules and ask him to change the subject. This was never necessary twice. No one wanted to anger a commissioner and prejudice his case.

I did not apply the same ex parte rules to relationships with officials of municipal systems, even though they were heavily involved in most of the cases on the FPC electric power agenda. They had been my friends during the TVA days and I did not want to give them up. The Federal Power Act is oriented toward the wholesale purchasers, mostly the municipals and the coops. Furthermore, I knew that the public's perception was that only fraternization with the investor-owned utility officials was improper. None of these reasons, however, will stand up under analysis.

At one time, the FPC issued bound volumes of reports of its decisions, but the practice of issuing such reports had lapsed. Each lawyer, for the commission or private parties, found it necessary to collect and maintain his own file of decisions. This might be possible for the old hands, but for a newcomer to build up such a file was nearly impossible. I instituted a new series of volumes, covering the current periods, and picking up the decisions for previous years at the same time. This work took some money and manpower (two assiduous and talented women); it was worth it, however, because it enabled the commissioners themselves, and counsel both inside and outside, to review quickly the history and issues of particular cases.

When I first came to the commission, I found that the commissioners, including the old hands, were not acquainted with the physical properties of the electric or gas industries. After more than a score of years working for the TVA, I had reasonable familiarity with electric power properties—the dams, steam plants, control rooms, transmission lines, transformer stations, and the rest. I had virtually no acquaintance, though, with the properties on the gas side. The other new appointees had little familiarity with the properties of either industry. Much of our work assumed that we had technical expertise, for which a knowledge on the ground of how the two industries appeared and functioned was essential. I arranged for a series of orientation visits to both electric and gas systems. On the electric side, these included both examples of public and investor-owned systems. On the gas side, they covered exploration, the laying of both underwater and underground pipelines, compressor stations, distribution systems, and the control rooms.

The visits were eye-openers for us all, and the companies were delighted by the evident desire of the commissioners to inform themselves. We paid our own way (at government expense and with public transportation), except where public transportation was not available, as from shore to a pipe-laying ship in the Gulf of Mexico. I might add that Kuykendall refused to go on such trips because he thought they might stir up criticism.

Personnel Changes

It is fundamental in trying to turn an agency around that some changes must be made in the people running the agency, not only the commissioners, but also among the top career people. Effecting staff changes can be traumatic, but this did not prove to be a problem at the FPC. The great number of employees were enthusiastically supportive, and there were many able people among them. It had been unpleasant to work in an agency that was criticized for mismanagement. Little had been done by the earlier commissioners to enlist their cooperation and support. They were eager to take part in the transformation of the FPC. There were only a few individuals—ones with the greatest potential for slowing or blocking the changes in policy and procedures—who left involuntarily.

Replacements for some positions were hard to find, but I was lucky. The word got around that the FPC had become an interesting place to work. For some jobs, particularly at lower levels, the FPC had more first-rate applicants than it could handle—idealistic young people who wanted to play a role in an agency with leadership they thought they could trust. I was in competition with academia and the large New York and Washington law firms for outstanding lawyers. Somehow I found out that Ralph Spritzer, assistant to the U.S. solicitor general, might be willing to consider a move. When I talked with him, I was tremendously impressed and made him an offer, which he accepted. His association with the FPC immediately raised the standing of the FPC's general counsel; Spritzer's appointment also gave notice that there would be no easy wins in litigation with the FPC. He was invaluable in the transition period, not only in his advice to the commission, but also in setting new standards of craftsmanship for the FPC lawyers.[3]

For the division of power post, I was able, after a long search, to recruit Stewart Brown, an engineer in the Bureau of Reclamation in the Interior Department, an expert on power engineering matters. He was totally dedicated, arriving at work early and leaving late, carrying a heavy briefcase. Since the power area was one in which I claimed some expertise, I probably worked more closely with him than with anyone else, except for my executive assistant, Dave Freeman. The division of power was a shell when Brown came, except for hydro licensing. Under his leadership, it filled out to the dimensions required to carry out all its statutory responsibilities. At the beginning, there were no rate experts in the division of power, although, strangely, four electric rate men were located in the gas division, where they had little to do because there were so few electric rate cases. These men were transferred to Brown's division.[4]

At the TVA, I had become accustomed to working with economists. I enjoyed their company and profited from their perspective on the TVA's problems. I did not always agree with them—they tended to be absolutists in pursuing their theories, without regard for their effect on people or the stability of institutions. They also had an innocent faith in economic models and computer printouts even when they were based on faulty data or logic. Still, we were in a position to exert a considerable influence on the energy marketplace, and I knew that the FPC needed an economics division. The head of this office was the hardest slot to fill. I must have interviewed fifteen or twenty people. I remember offering the position to Alfred E. Kahn of Cornell, but he turned it down. Finally, I found a suitable person, Haskell Wald, who organized the FPC's new office of economics and worked for the commission for many years.[5]

Two key appointments were my personal assistants. Dave Freeman, serving on the staff of the TVA general counsel, had approached me in Nashville to ask for the appointment. Freeman had a broad assignment, to follow up on my decisions and the commission's orders, and to be sure that there was movement in the Permian Basin natural gas rate case and the National Power Survey, both discussed later. His functions required him to serve as a liaison with key staff people; I also used him as a sounding board to test out some of my ideas, both for action and for speeches. He proved his worth in these roles.

My other special assistant was Donald Young, a holdover from the Kuykendall days. He was familiar with the details of the FPC's work. He was perceptive, retentive, and serious, but with a good sense of humor. His principal assignment was to brief me as needed on the sometimes complex background and issues in proceedings before the commission, including a regular briefing on the day before the weekly commission meetings. Furthermore, he prepared notes on these matters for my use. He was ideal for the purpose, and I came to like and trust him.

CHAPTER 17

The Dynamics of Regulation under the New Commission

Each government agency operates under a complex statute. Its goals are often diffuse and difficult—sometimes impossible—to attain. They are circumscribed by a multiplicity of legislative and executive controls, including personnel and money. Collegial agencies, those headed by boards or commissions, have a special problem. Their fate can be sealed by appointments, not merely of the chairman, but equally of the other members. If the chairman is to have more than a fighting chance to move the agency along, he must have the president's support on appointments.

I had no commitment from President Kennedy to seek my advice, much less my consent, on intended nominees. I made clear to Ralph Dungan and Lee White, however, that the opportunity to reform and improve FPC performance could be lost if the individuals appointed by the president had a different conception of the public interest than his and mine. Dungan eventually told me that the president would endeavor to assure a like-minded majority. As a practical matter, I was given an opportunity to interview all proposed appointees and discuss them with Dungan. Some I was happy to endorse. Others I viewed with serious reservations, but not acute alarm, because the appointments did not disturb majority control. I understood, of course, that influential senators, whose support Kennedy needed for both domestic and international initiatives, were pressuring him to appoint people they recommended. I was reluctant to press FPC matters when the action required from the president collided with much graver questions on the national agenda. I said much the same thing in an interview with David Halberstam while I was in office. As it turned out, the appointments about which I had reservations ended up being helpful, rather than otherwise, because (I like to think) they were co-opted by the dynamics of an agency with a sense of mission and an accumulating record of performance.[1]

My Fellow Commissioners

Morgan may well have been promised an appointment before I was recruited. I did not interview him in advance, and was told by the White House staff that I was sure to like him. His appointment was strongly recommended by Senator Morse. Morse was capable of persuasive eloquence, but was also known for his headstrong speech, his headline hunting, and his unreliable conduct. Morgan was endorsed by the other Oregon senator, Maurine Neuberger, a liberal on all issues. I found it difficult from the first to get along with Morgan, although I tried. He had a good mind and could be good company when he chose, but he was vain, and despite large claims for his abilities, he did little to improve the performance of the FPC.[2]

I had no problem with either the Ross or Black appointments. They were lawyers, and good ones, liberals in the classic liberal tradition. Ross was a Republican, a protégé of Senator Aiken, perhaps the most liberal of all the senators on the Republican side. Ross's primary interest was environmental protection. His positions on that issue were well ahead of his time. He was also a public power man, though not necessarily anti-utility; in any proceeding, he favored the lowest gas or electricity rates that could be justified from the record. He was politically ambitious and uncomfortable in a background role. A sensitive and stubborn man, it was necessary for me to show him respect and attention, which was not difficult on most votes because our views were much the same—or his viewpoint was not unreasonable and could be assimilated with mine. Black was sponsored by Senator Henry Jackson, an old friend of mine. Black was the youngest of the appointees. With Ross, we could predict two sure votes for the "liberal" result in resolving almost any problem.[3]

O'Connor was of a different stripe. He was serving in the Interior Department as chairman of an oil industry advisory board. Trained as an accountant, he had worked for oil companies in Texas and brought with him an inside knowledge of the industry that the other commissioners lacked. He was sponsored by Speaker Sam Rayburn. When I expressed misgivings to Ralph Dungan about appointing someone from the oil industry to the commission, Dungan told me the president believed he must accept Rayburn's choice. I need not have worried. O'Connor was an able, independent, and fair-minded man; his positions were always based on principle. He was not contentious and took pride in the work of the FPC. I came to respect him greatly.[4]

When Kuykendall resigned, he was replaced by Harold C. Woodward, the choice of Senator Everett Dirksen, a Republican leader whose law firm in Pekin, Illinois, was retained by the Tennessee Natural Gas Company. He was under Dirksen's patronage as an officeholder in Illinois. I tried to head off this appointment, but Dirksen was regarded by the White House as the most powerful person in the Senate. The success of Kennedy's bipartisan foreign policy depended on him. Dungan pointed out that with the votes of Ross and Black, I could still have a liberal majority when I wanted it. It was agree or resign. I agreed. Again, my fears were overtaken by a pleasant reality. Woodward was a team player, with few settled opinions on energy matters. He was quiet, undemonstrative, and easy to like. To him, the post was a reward for his work for Dirksen in Illinois. After he came to know me, and had witnessed some of my difficulty in forming a majority on some controversial questions, he volunteered privately that I could count on his vote on any issue unless it conflicted with the desires of Dirksen. This was comforting. His dissents were few and did not change the result in any case; indeed, he made no effort to persuade the rest of us. In reality, I had more problems with the strong-minded and assertive liberals than with the considerate and respectful conservatives on the commission.[5]

Woodward was succeeded by Carl Bagge, a lawyer who had been a part of the Dirksen political machine. Temperamentally, he was the polar opposite of Woodward. Young, vigorous, and ebullient, he had strong opinions on everything, which he gladly shared. Of Swedish extraction, he was fond of Swedish food, stories, and songs. To his credit, he proved a disappointment to his sponsoring senator. On an important issue, he voted against Dirksen's advice (or instructions). He told the rest of us he knew he was in for trouble, but he had to vote his conscience. He was renominated, but Dirksen held up Senate action for months. I do not know what went on behind the scenes, but eventually Bagge's nomination was approved, without overt opposition by Dirksen.[6]

It was evident from the performance of Dirksen's two men that he received little benefit from placing Woodward and Bagge on the commission—perfunctory compliance from Woodward and intractability from Bagge. This history may prove something about the spirit of independence of American officeholders. Or, it might demonstrate the inherent difficulty of assuring that an official, once bought, will stay bought, or the tendency of an organization in a dynamic phase to bring out the best in its workers. Perhaps it only shows that Dirksen made a couple of poor choices for his purposes.

Reviewing and Deciding Cases

I believe that commission meetings were held bi-weekly under Kuykendall. We met weekly during my tenure, partly because we not only used the meetings for voting, but also for time-consuming discussions to hammer out decisions on cases and policies. The commission met on Tuesdays, usually for the whole day. An agenda with supporting documents was circulated among the commissioners and key staff by Friday of the previous week. We had a rule that no item could be considered at a meeting unless it was on the agenda. The rule was ironclad, but the ironcladding was malleable, and it was a rare meeting that we did not consider several such items. In practice, if a matter was ready for decision, and delay would be costly to the industry or consumers, we waived the rule. The rule did put pressure on staff to bring up items in an orderly way. It also gave commissioners an excuse for grumbling, an exercise in which all commissioners could happily join. The watchword of the agency was to expedite the regulatory process. If the commission tolerated unnecessary delays, the staff's momentum would be affected. When an item was more difficult than staff assumed, or when a commissioner wanted to study it in more detail, it was put over to a following meeting. Most meetings were even-tempered. There was much joshing, and the staff seemed to enjoy this comic relief in meetings that were long, serious, and frequently tedious.

On the weekends, I usually reviewed two large loose-leaf volumes, one each for gas and electric items, containing all the agenda items and key supporting materials. The non-routine items were accompanied by a brief explanation by Don Young. On the most difficult matters, I made time at the office to study the principal briefs and other papers. On other matters, I spent four to six hours on most Saturdays and Sundays, making my own notes and listing the questions in my mind for discussion with Young, Freeman, or staff members. The other commissioners received the same material at the same time I did, Young's notes excepted.

Sometimes, against my better judgment, but feeling the need to move on with the agenda, I pushed the commission to move faster. Occasionally, I tried to cut off a commissioner at the third or fourth repetition of an argument. I discouraged digressions. Staff displays of learning were also discouraged if they were not brief and to the point. Above all, I never refused a personal request from a commissioner if I could help it, whether to reset a meeting, to advance or postpone (within limits) the consideration of an item, to use their proffered phraseology in writing opinions, to give full credit for their contribution in solving our problems, or to accede to their requests for preferment of one kind or another.

The FPC came to be distinguished among regulatory agencies by the number of big cases it handled. In a long career, a U.S. District Court judge might not handle any case that involved as much money or had as much impact on consumers, industry, or the country's resources, as the FPC did. We handled them by the dozen. Even without foot-dragging, these big cases took months, sometimes years, to try because of the large number of interests, parties, and issues involved. As we built up our ALJ staff, other agencies, especially the FCC and the AEC, began to lean on us, recruiting some of our ALJs or borrowing one for a particularly difficult case.

At early meetings, Kuykendall would occasionally refer to what he called the "X agenda," and once or twice he suggested transferring some troublesome item from our regular agenda to the "X agenda." The "X agenda" turned out to be the commission's black hole. Items placed on this agenda were never heard of again, although they remained on the commission's docket (some were a decade old). The commission did not review items on the "X agenda." I saw no point to a special dumping ground for toxic matters that the commission did not know how to handle. At the first opportunity, I asked that these items be placed on the regular agenda. Some were indeed difficult, important, and overripe. The commission took up one or more of the items at each meeting. Some involved aborted rulemakings. These were either terminated or put on a schedule that would lead to the issuance or denial of a rule. Others were contested matters that had not been pressed by the parties, although some involved significant issues. These we tested for signs of life. If the parties had forgotten the issue or settled the matter themselves, the case was terminated. Otherwise, it was heard and decided. Some of the rulemakings were in limbo because they required hard decisions by the commission on how or whether to proceed. These were the toughest cases.[7]

As an example, hundreds of dams, some fairly large, were operating without licenses. Frequently, they involved complicated (and contested) jurisdictional questions. There was ample room for contention. Should we pursue all these cases? Our licensing staff was small and already busy. Normally, initial licenses were issued for fifty years. Should the illegal operation of these plants be blessed by giving them all fifty-year licenses, beginning currently? If a lesser term, how long and beginning when? Some of the hydro plants were owned by small industries or municipalities. Should they be treated differently? When most of the plants had been built, the Supreme Court had not yet made clear the navigability criteria on which jurisdiction depended. A special problem was the FPC safety responsibility. Even a small dam immediately above a town could do great damage and cost many lives if it failed. We could not close our eyes to the licensing problem, if for no other reason than insuring dam safety.

A staff committee was appointed to study the problem and make recommendations. The committee came up with a set of standards and a program to determine the status of those plants whose owners resisted our claims of jurisdiction. With small changes, the staff recommendations were adopted. Most dam owners acquiesced gracefully when they saw that the FPC meant business, and that they could be subject to severe penalties if they continued to operate without a license. Over a period of several years, most of the problems were resolved and most of the dams within our jurisdiction (a few were found to be non-jurisdictional) were licensed; this subjected them to our requirements for safety inspections, public access, and recreational development.

During my tenure, there were no "sunshine" laws like the ones that now prohibit private meetings among members of federal regulatory agencies, or require that all substantive meetings be open to the public. Our meetings were not open to the public. It was thus possible to use meetings as a forum to test ideas, to help reach individual decisions, to change a decision when called for, to persuade others to one's point of view, or to go over the wording of a decision for both substance and style. None of this is likely in a public forum. For example, what commissioner will say, before God and all the world, "Here's a possible way out of the dilemma I haven't thought through, but suppose . . ."; "On reflection, I withdraw the stupid proposal I made last week"; "I must have dozed off, would you please repeat the point you just made?"; "Would you say that again in words of one syllable?"; "Let's try harder; we can't let ourselves get stymied on this one"; "Company [or commission] X tried that road, and look at what happened"; "Suppose we split the difference so that a majority can concur in the result even if not on the rationale?" It seems to me that commissioners should be free to use all of these and other approaches in deciding their cases and solving their problems, without fear that a spontaneous expression will be taken out of context and create a false impression.

Inevitably, there is now an element of posturing for the audience, at least in the negative sense of withholding views that might lead to vigorous challenge. The opportunities for compromise and persuasion are greatly reduced, positions are hardened before they are thought through, and confrontations are more likely. There is less likelihood of pointed probing by commissioners on facts or assumptions. Joking remarks that would be brushed off in good humor in a closed meeting may give offense in an open forum. For the most part, sunshine meetings merely record and explain positions that have been arrived at individually in private. Rather than being decided on the spot, there is much more likelihood that tough questions will be referred back to staff for further work. The irony is that the meetings have become so dull that reporters for the general press do not attend.

My own style requires more spontaneity, more informality, more opportunities to release tensions, and more risk-taking in exchanges with other commissioners than are possible under sunshine laws. It is better, on balance, that an occasional miscreant should abuse his position than that the vitality and viability of the whole system be eroded by treating the commissioners as so undeserving of trust that proctors must attend their every meeting.

It will probably have registered on anyone who has read this far that philosophically I am a gradualist. I cannot say I believe the world grows better all the time, but I do believe that individuals can affect the course of events for the better, as compared with standing to one side. They have an obligation to try, all the more if they are in positions of influence. I believe also that the surest course is to move a step at a time, at a steady pace, to preserve what we have until we find a better way, instead of gambling on the unknown.

CHAPTER 18

How Life Was Brought to FPC Electric Power Regulation

What, specifically, did I intend to accomplish in my new job? I had no formal agenda. I knew that somehow I must vitalize federal regulation of power transactions—not revitalize, because there was no period when it had been effective. I knew from my TVA experience that much of the electric power industry was not operating at a reasonable level of efficiency, considering the advances in power technology. Coordination among companies was inadequate. Many and stronger transmission ties were needed. The gap between the best and the average company's performance was far too great. The industry was bedeviled and distracted by the hundred-year war between the investor-owned and publicly-owned systems. Many important decisions on both sides were made, not on their economic merits, but as gambits in that war. Gas regulation was also a disaster. These three lines of effort—establishing the FPC's role in power regulation, improving the performance of the electric power industry, and conquering the gas regulation jungle—were my starting objectives. I had no blueprint for accomplishing any of them, but I seized opportunities as they arose.

Did I worry that in the future the economists at the University of Chicago, or Stanford, or the Massachusetts Institute of Technology might think that it would be better for the country if both electric power and natural gas regulation had been scrapped and the FPC had been relegated to its original mission of licensing hydro plants? I did not. The era of the radical right-wing economists, who believe in applying free market principles even where no free market exists or is possible, had not yet begun. It never occurred to me that I was not working in a good cause. I believed that regulation of both industries was desirable and in the public interest. Regulation, of course, had to recognize that managers and investors must be fairly compensated, with rewards for superior performance. Congress had provided for electric power and natural gas regulation, the Supreme Court was supportive, and the public looked to improved regulation to protect it from industry abuses, shortages, and high prices. I believed that it was my job to carry out the mandates of Congress and the Supreme Court, and to help improve the quality of federal regulation.

Almost as soon as I got on the job in Washington, I met with the senior power staff and a few lawyers who were familiar with the court decisions in electric power cases. My purpose was to find out why the Federal Power Act provisions for the regulation of electric rates in wholesale transactions, and other crucial provisions introduced in the 1935 amendments, had not been put into effect. It turned out that the basic reason was a failure to develop a legal theory that would support jurisdiction. This was a strange reason considering that only a strained interpretation of the act could deny that it placed the investor-owned companies (some 80 percent of all electric suppliers) engaged in interstate power transactions under FPC rate jurisdiction. The prevailing FPC technique for determining whether a transaction was interstate, and thus subject to jurisdiction, was to trace the power flows in order to ascertain if the load was serving plants outside the state where it was generated. Since most companies operated in a single state, and had generating plants within that state, they could claim that any given delivery was from a source within the same state, even if the company in question was part of an interconnected network covering several states.

I was not an engineer, but I knew enough about power system technology to be aware that the test was worse than ridiculous; it was fraudulent. Electricity is not a commodity, like gas, where tracing flows in an interstate system is difficult enough. Electricity is "transmitted" by arranging the electromagnetic forces on a transmission grid to produce the desired deliveries. There is no way of determining the paths of the impulses that are delivered. Even the largest investor-owned system in the country, the American Electric Power Company, could boast of the efficiency of its integrated seven-state system and deny, for purposes of the Federal Power Act, that the energy moved in interstate commerce.

In locating plants, the inherent benefits of a particular site were often overlooked; instead, utilities were more concerned with being able to claim immunity from FPC jurisdiction. The stakes were large, estimated by engineers at many billions of dollars a year. The uncertainties whether the federal government could or would claim jurisdiction inhibited the construction of new ties and the optimum use of existing ones. They also discouraged joint planning and the formation of regional pools. The question had to be decided, one way or the other, and federal authority was by far the better answer. Otherwise, each of the states involved in a multi-state pool would seek a disproportionate share of the benefits.

I was so convinced in my own mind that the courts would rule that interconnection across state lines with switches closed (that is, when the system is electrically, as well as physically, connected) constituted a jurisdictional operation, that I posed to the engineers the next question. Assuming that I was right, what companies were jurisdictional on the basis of my interpretation of the law? The engineers said it would take extensive studies to answer my question. I pressed the point—did they have any idea which companies were physically interconnected systems? Had no one even speculated on this crucial question? Finally, one engineer confessed that there was a piece of paper listing some of the companies in a tentative sort of way. I demanded to see the list. It proved to be complete, up-to-date, and the product of careful analysis. I did not press for the reason staff had withheld the document, but presumed that in prior administrations the staff did not want to make waves for fear of creating trouble for themselves. I was too pleased to have the list to allow myself to be distracted from putting it to use.

It is difficult even in retrospect to explain adequately what the officers of these companies feared. The history of appointments to the FPC of men with strong anti-utility views made many executives apprehensive. The Federal Power Act itself, truth to tell, is skewed in an anti-utility way by exempting public systems from regulation, while at the same time making them the beneficiaries of the commission's efforts. Utility officials might well have considered that the staff, recruited in earlier days, was predisposed against them, a view not entirely fanciful. At the time, it was a common assumption that state regulators would be less coercive than the Washington agencies, more generous in rate cases, and subject to local political pressures. In the main, however, I believe it was the fear of Washington, of losing some of their freedom and initiative, and of being mired in a bureaucratic swamp. This fear was real but exaggerated. The need for a better use of existing facilities, and for the addition of new facilities, required federal regulation. Governors and state legislatures, however, were more likely than the president or Congress to adopt extreme and unfair measures [toward utilities] in response to consumer protests. It is ironic that a quarter century after the FPC established jurisdiction against strong utility opposition, many of them were trying to restructure their organizations (by placing generating facilities in a separate company) in order to bring rates within the FPC's regulatory jurisdiction, rather than state regulation.

As it happened, a suitable vehicle for a definitive legal test was then working its way through the courts, the so-called Colton case. The city of Colton, California, which owned a distribution system that purchased its supply at wholesale from the Southern California Edison Company, petitioned the FPC on May 9, 1958, to assert jurisdiction (the city was dissatisfied with a rate increase filed by Southern California Edison and authorized by the California Public Utilities Commission). The FPC did, in fact, assert jurisdiction, using a form of tracing to justify its decision. Southern California Edison then petitioned the U.S. Court of Appeals for the Ninth Circuit to deny FPC jurisdiction. Ralph Spritzer, FPC general counsel, argued the case, but the court, one judge dissenting, set aside the FPC decision on November 15, 1962. The FPC carried the case to the U.S. Supreme Court, which unanimously endorsed Spritzer's contention that sales at wholesale from interstate systems are essentially national.[1]

It was important to introduce effective federal regulation quickly. The late 1950s and early 1960s represented a period of rapid technological improvement, reflected in increased efficiencies and relatively stable prices. It seemed clear that customers at wholesale were not being given the benefit of the reduced costs, and that there was much room for rate reductions once the jurisdiction of the commission was established. If the FPC proceeded in the normal way by starting one or more new test cases, and resting on its oars while those cases wended their way through the courts, practical improvement might not arrive until long after my term of office expired. I suggested a shortcut. The commission should make public our list of jurisdictional utilities, require these companies to file their rates—that is, to submit to our jurisdiction—or to show cause why on the facts and the law they were not operating interstate power systems. The result was to place the burden and risk of noncompliance on the companies. For example, not all states regulated the issuance of securities, and few states regulated the issuance of all types of securities (FPC had control under its statute over almost all security issues that were not subject to state approval). The statute was sufficiently clear, or perhaps from the utility viewpoint was sufficiently ambiguous, that no utility counsel, bond counsel, or counsel for the banks could advise clients or bond buyers that the companies we named were non-jurisdictional beyond reasonable question.

In the meantime, the lawyers and engineers had been refining the legal theory. An FPC electrical engineer, Joseph Jessel, was especially helpful. In his technically sound formulation, the interstate systems—that is, all systems that crossed state lines unless they were purely distribution systems—were locked together by electromagnetic forces and functioned as a single system. It was a formulation that ultimately found ready acceptance in the courts and doomed the efforts to evade FPC jurisdiction by companies engaged in interstate transactions at wholesale. A number of small companies were granted exemptions on the facts (for example, when the only interconnection was over a low-voltage line used for retail service). Other companies, small in number but large in size, contested jurisdiction, all successfully. Despite the 1962 utility victory in the court of appeals, most of the utilities had little confidence in the litigation. They cooperated long before the Supreme Court disposed of the issue in 1964. No penalties were imposed on any company for delaying compliance pending the final outcome in the Colton case. It was not altogether their fault, for one thing. Lax administration had lulled them into a false position. Moreover, my plans for the industry rested in part on a statutory provision for encouraging voluntary arrangements among utilities, a goal that could only be achieved by cooperation and mutual trust. Efforts at punishment would lead nowhere. A few companies decided to litigate; the others, accounting for most of the industry, dutifully filed, with the knowledge that if the courts upheld the litigating companies, they would enjoy the same exemption.

The case of Detroit Edison was particularly interesting. Its operations were tied to another large Michigan electric company, Consumers Power Company. Neither company had been willing to interconnect with its neighbors in other states. Their joint network, however, was interconnected with the Ontario hydropower system in Canada at its western end; the Ontario system was tied in with the New York electric systems in the east. With all switches closed, the Michigan companies could exchange energy freely with the New York companies and with other systems in the United States. Nevertheless, the Detroit company's chairman, Walker Cisler, solemnly assured the FPC that all switches were never simultaneously closed. Detroit Edison's agreements with Ontario Hydro prevented simultaneous transactions with U.S. companies at both ends of the Ontario system, so that power could not flow from Michigan to New York through Ontario, or from New York to Michigan. This meant that at all times at least one set of switches on the Canadian system would be open, according to Cisler. The case for FPC jurisdiction would have been debatable if this had been true. It was not.[2]

Once the FPC list was published, I received a letter from the senior senator from Michigan, James Couzens, asking on what basis we could possibly claim jurisdiction over Detroit Edison, considering the facts supplied by the company. Couzens was widely respected for his courage and integrity. With my reply, I enclosed a clipping from a recent issue of the *Electrical World* in which Cisler was quoted as taking credit for having averted a power shortage in New York City by supplying large amounts of power through the Ontario connection, something that was possible only if both sets of switches were closed. Couzens wrote me no more on behalf of Detroit Edison.[3]

Shortly afterward, Cisler offered a settlement proposal: Consumers Power would build an interconnection with a neighbor in another state, whereupon both Michigan companies would accept jurisdiction. The proposal would postpone jurisdiction for a couple of years, but it would permit the relatively early addition of the Michigan companies to the integrated pool of the Eastern and Midwestern companies, with substantial economies and benefits for all concerned. It would also save the delay and expense of litigation.

The assertion and acceptance of FPC jurisdiction over interstate transactions at wholesale—covering all the major systems in the country, except for

Texas—had many benefits. No longer was fear of inviting federal jurisdiction a barrier to system planning on an efficiency basis. Now the FPC could encourage more and stronger interconnections, and the companies could build them, assured that they would be treated on their merits. Much was done in this direction during my tenure. Equally important, the wholesale customers, especially the municipalities and cooperatives, could now count on reliability of supply and on an effective forum to decide disputes on rates and service. The municipalities and cooperatives were then engaged in enterprises to build their own generation plants because of the lack of rate protection or continuity of service provided by the private systems.

The newly acquired security in rates and supply should have preempted plans by the buyers to build their own generating plants and to discontinue power purchases, but it did not work out that way. With federal subsidies, by way of exemption of bondholders from federal taxes, they had access to cheap money; in the case of the cooperatives, they could receive direct government loans at 2 percent interest. With such subsidies, even far less efficient plants than those of the utilities were "cheaper." It turned out that the municipalities and cooperatives were less interested in consumer savings than in building their own generation and transmission empires. They believed, I suppose, that in time they would attain the same efficiencies as investor-owned systems. I did not think the federal government should lend money or sacrifice revenues to finance new power plants when lower-cost power was available for purchase on secure terms. The U.S. could have saved billions of dollars, and both consumers and taxpayers would have benefited. I made this point to the Bureau of the Budget; they agreed with me, but my protests never got anywhere. The rural cooperative lobby in Congress was too strong to permit turning off the tap on federal money, and the municipal income tax exemption was too deeply rooted for successful challenge.[4]

Another benefit of effective jurisdiction, which did not seem important at the time, has been the ability to settle disputes among the utilities themselves. The most notable examples were the disputes among companies within a holding company system, or with the respective state commissions that regulate them, as to the allocation between companies, each operating in a single state, of the costs of nuclear plants, which (later data showed) proved to have the highest costs. The state utility commissions entered the battles on behalf of the consumers in their respective states. The FPC and its successor agency, the Federal Energy Regulatory Commission [FERC], have proved indispensable in settling these highly emotional disputes, in which, for political reasons, none of the companies or commissions involved could afford to give ground.

CHAPTER 19

New Programs in Electric Power Regulation and the Northeast Power Failure

Once the FPC had established its jurisdiction on a solid legal footing, it was up to us to exercise that jurisdiction in an organized and aggressive way. Teams of lawyers, engineers, and accountants examined the books and reports of all the major companies. In many, if not most, cases the utilities were not adequately sharing their savings with their (wholesale) customers. The FPC initiated proceedings against all such companies. In a few instances, it was necessary to conduct a full-fledged formal case, but most were settled without litigation; rate reductions became the order of the day. Each wholesale rate reduction was publicized in the affected local area so that it could not be overlooked by the state commissions or the press. These reductions were usually translated by the state utility commissions into rate reductions for retail customers.

Rate Reductions

My tenure lasted long enough to cover a succession of reductions, and this raised a basic question of regulatory philosophy. Were we primarily interested in holding down a company's profits or in bringing about cost reductions that would be passed on to its customers in the form of lower rates? When the staff got in the swing of the rate reduction program, it wanted to trace earnings and bring rate cases every six months, or every year if the earnings exceeded a certain amount. One or two of the commissioners were of the same view. The trouble with this approach, to my mind, was that it provided no incentive to reduce costs. If the company was to lose the benefit of savings, why should it explore ways to cut costs? I believed that the rate reductions should be phased in at two-year intervals, to provide periods of increased earnings, and thus a continuous incentive to reduce costs. At the end of that time, the savings were likely to be much larger than if semi-annual or annual rate reductions were required. I argued that to get the best performance from a utility, something should be left in the pot for the stockholders.

The argument was carried on at several commission meetings. The big-stick side figured that the effect of incentives was uncertain, and that we should give maximum current benefits to consumers (and take greater credit for our work). In truth, it would take a while to prove the benefits of delayed rewards for improved performance. Those on the carrot side considered this approach shortsighted. Their primary aim was to stimulate innovations in cost cutting. As I recall, we never reached a consensus in principle, and we made ad hoc decisions without setting definite standards, but the staff was impressed by the advantages of the long view.

Recreational Use of Hydro Reservoirs

Hydro projects were not utilized much for recreation. Development at reservoirs built by private interests was almost non-existent. Reservoirs behind the dams licensed by the United States were also neglected. I saw no reason why this situation should continue.

The solution was within the commission's jurisdiction. The Federal Power Act granted the FPC broad discretion in framing the terms of hydro licenses. The FPC, a year or two after I took office, adopted a regulation that required every application for a license, or for a license extension or amendment, to include an exhibit (called "exhibit R") that would set forth a comprehensive public recreational plan, including a commitment to open access. Staff recreation experts, employed for the purpose, analyzed the proposals for adequacy and comprehensiveness. Local organizations contributed their comments. In the end, agreement was usually reached. Experience has shown that these recreational programs are popular with local communities and bring a public relations benefit to the utilities. Public use of licensed reservoirs is now all but universal, a national resource that costs taxpayers nothing.[1]

Dam Safety

Dams require constant surveillance and maintenance. The FPC recognized its responsibility to insure that dams were safe. When I came on board, though, it had done little in this area. I knew that at the TVA safety was a primary concern. Failure of a large dam could send the equivalent of a tidal wave down the river, destroying everything in its path, even other dams. The FPC engineers were insistent on safe designs, in accordance with prevailing engineering standards. Maintenance, however, was equally important, and the condition of most of the dams was unknown.

In my early months on the job I gave little, if any, attention to dam safety. I was fully occupied with other matters. One day, though, the papers reported a catastrophic dam failure in Italy that killed hundreds of people and destroyed a great deal of property. I suddenly awakened to the risk of a similar disaster in the U.S. involving a licensed project. With no further delay, I arranged for a briefing of the full commission by the chief of the bureau of power and his staff. We found an alarming situation.[2]

The FPC did have a safety program, but it turned out to be worse than no program at all because it provided an illusion of safety. Only a few engineers were assigned to the safety function. They visited some of the projects for brief inspections at long intervals. Many projects had not received even cursory examinations. The problem was one of money. To do an adequate job would require scores, if not hundreds, of engineers, geologists, and seismologists. The costs of hiring these individuals would have dwarfed our budget for the bureau of power. It was beyond any amount the Congress was likely to appropriate.

Of course, it might not have been necessary to carry out inspections with our own staff. At my suggestion, the commission had recently agreed to accept audits made by the utilities and certified by any of the well-known CPA firms, thus enabling the commission's limited auditing staff to focus on suspect or controversial areas. Could we not accept reports of engineering firms, supported as necessary by geologists and seismologists? These reports and the underlying working papers would call attention to a wide range of safety issues. Reliance on independent engineering firms would make available the expertise of the whole country, including consultants drawn from academia, a fount of talent that the commission could not equal.

The staff believed that this proposal was too simple to work, but were soon persuaded of its merits. The proposal was reduced to a formal rule that called for annual dam safety reports by the utilities themselves, as well as comprehensive inspections by independent engineering firms at least every five years. If the reports indicated a problem, the commission would conduct an immediate investigation. The commission was authorized to retain consultants, unaffiliated with either the licensee or the independent engineering firm, at the expense of the licensee.

Headwater Benefits and License Fees

Two related FPC functions had almost totally lapsed when I joined the commission. The first concerned the requirement that investor-owned companies had to compensate the federal government for benefits they received from being downstream of U.S. dams. The second was billing for fees in connection with the processing of hydro license applications.

Dams built for flood control purposes are usually located at relatively high elevations between narrow canyon walls. These dams, controlling large volumes of water, are also likely to be efficient power producers, since the amount of power capacity is a function of water volume and "head," the latter being the distance the water falls before it hits the blades of the turbines. Below such high dams, on the tributary itself or on the main stem below the point at which it is joined by the tributary, there are frequently other dams that generate power. The high dam is likely to contribute a number of benefits to these dams. By regulating the flow of water in the flood season—when there is more water than could be used in the generators of the lower dams—and by releasing the water in dry periods when the generators are not fully loaded, the high dam makes additional power generation possible from the lower dams.

Aware of these advantages, collectively known as headwater benefits, Congress made provision for compensation to the United States by the private power companies that received them. The FPC was given the responsibility for determining the dollar value of the headwater benefits and for collecting the amounts. The total received through the 1961 fiscal year was approximately $600,000; that just about equaled the administrative costs associated with its collection.

The new commissioners were appalled by this record. The benefits to privately owned power projects from the public investment in upstream storage projects were very large, but the commission had been unable to make timely determinations of the amounts. Many millions of dollars that the Treasury Department was entitled to were lost forever. The failure to design an effective compensation process was of a piece with the FPC's inability to develop a workable method to determine which electric utilities came under its jurisdiction and its failure to determine the field price of gas. On headwater benefits, it was soon clear that annual determinations on a dam-by-dam basis were ineffectual, and that no amount of detailed tinkering would salvage this [confusing] approach. What was needed was a formulaic approach applied to a limited number of variables.

At the commission's direction, the staff provided such a formula, one that dealt rough but prompt justice. It was inserted in new licenses so that the licensees had a contractual commitment to pay a fee that was relatively easy to calculate. The results were almost instantaneous. I testified in the 1964 House appropriations hearings that collections were $168,000 in fiscal year 1962; they were projected to be $1,700,000 in 1963, and $4,000,000 in 1964. Indeed, collections increased from year to year as new projects were licensed and the unit value of power benefits rose.

The problem of licensing fees was easier to solve. Section 10(e) of the Federal Water Power Act (1920) required that each licensee pay annual fees to the United States for the cost of license administration. So far as I can recall, these determinations were timely, and the annual amounts collected were about $1 million by the early 1960s. I testified during the 1963 appropriations hearings that "in recent years charges . . . have not been sufficient to cover the licensing function." I also stated that the commission was taking action to increase the assessments to assure "that annual charges hereafter fully cover the appropriate administrative costs as intended by Congress." For 1963, receipts were estimated at $1,389,000. They increased annually thereafter in step with the increases in the commission's own costs.[3]

The Abortive Effort to Bring G&T Cooperatives under FPC Jurisdiction

Not all the FPC initiatives were successful. One was the attempt to apply the Federal Power Act to the so-called generation and transmission (G&T) cooperatives. These were the fifteen to twenty central cooperatives organized and owned by groups of distribution cooperatives, which generated and sold power to their owners. The G&Ts were tied into the interstate power networks, and there was no question that they sold power in interstate commerce. In form they were private enterprises, a matter of pride to themselves and the consumers of the distribution cooperatives. The Federal Power Act exempted systems owned by public entities, municipalities, counties, and states, but there was no express exemption for private systems organized in cooperative form. It appeared that on numerous occasions some of the G&Ts had filed their rates with the FPC, and that counsel for the Rural Electrification Administration (REA) had advised REA borrowers many years before that G&T cooperatives came under FPC jurisdiction if they were part of an interstate transmission network. Nevertheless, the FPC had never acted to assert jurisdiction, just as it had never made clear to which investor-owned companies its jurisdiction extended. We therefore initiated several test cases to determine the question of cooperative jurisdiction under a variety of factual situations. The events that followed provided an instructive example of what happens when an administrative agency confronts an interest that enjoys protection from influential members of Congress.[4]

The electric cooperative lobby is perhaps the strongest in Washington, rivaled only by the National Rifle Association and the American Association of Retired People. There are more than 900 distribution cooperatives, which serve several million members at retail, spread into almost every state. In addition to the protection of their guardian angels, the REA and the Department of Agriculture, the coops—both distributors and G&Ts—have organized their own national association: NRECA. The coops within a state are also organized informally by congressional districts. It is lobbying gospel for the coops in any state to meet regularly with their respective congressional representatives. The Washington staff of NRECA is ubiquitous in the halls of Congress when matters even remotely concerning the coops are up for consideration.

The coops have other important allies. On many matters of FPC regulation, for example, they form a common front with the American Public Power Association, which has about a thousand members, mostly local systems owned by municipalities, counties, and power districts, but including some state-operated and regional systems as well. On questions of jurisdiction an effort to regulate the coops was regarded as a possible precursor to a similar effort with respect to public systems, and as an attack on both, even though the reasons for exempting public systems do not apply to coops. The statutory exemption of the public systems was clear and explicit, and the FPC disavowed any intention to seek such jurisdiction. The cooperative movement had also attracted a variety of other sympathizers and helpmates who would have responded to a call for assistance from the "little fellows" without digging too deeply into the merits of the particular occasion.[5]

The weight of prospective opposition from these powerful groups would give pause to any official, and we paused long enough for a couple of spine-stiffening sessions. But our general counsel advised us that under the Federal Power Act, the commission had jurisdiction over the G&T coops engaged in the sale of electricity at wholesale. I, for one, did not see how we could avoid arriving at a formal conclusion on the question of jurisdiction, any more than we could waive jurisdiction over the investor-owned companies. It was not as if the G&Ts were negligible in size and activities. They were becoming a substantial factor in electric power supply, building generating plants of increasing size and hundreds of miles of high-voltage transmission lines. They were directly serving hundreds of distribution cooperatives and, through them, millions of consumers. It seemed to me that the FPC's effort to integrate the nation's power systems would have to include the G&T cooperatives. It was easier to defend ourselves against the charge of attempting too much rather than explain why we settled for too little.

The usual way to determine jurisdiction was not to rely on our own opinions, but to issue an order to show cause why we should not assert jurisdiction. We would then await the outcome of proceedings under the order before arriving at our own conclusion. In this way, the relevant parties made their case before an ALJ. Next, the ALJ would write an opinion on the basis of the law and the facts brought out in the hearing, briefs, and oral argument. Dissatisfied parties could appeal to the commission and seek rehearing if still aggrieved when the commission decided the case. If rehearing were denied, they could appeal to a United States Court of Appeals and, ultimately, seek relief in the United States Supreme Court. Accordingly, we issued orders to show cause in three cases in order to test the jurisdictional question in a variety of situations.

These proceedings never came to life. No sooner were they announced than the FPC was confronted by a barrage of opposition. Our actions were attacked as a power grab, or as a blind reliance on staff miscreants who had led us into grievous error. The REA Administrator, Norman Clapp, a good man and a personal friend, joined in the attack. The full lobbying resources of NRECA were brought into play, and they were formidable. Soon the attack moved to Congress. Bills for explicit exemption were introduced in both the House and the Senate. The Senate Appropriations Committee stated that none of the FPC's funds should be used to prosecute the coop cases. I was also invited to appear before the Senate Commerce Committee to testify on the bills to remove any FPC jurisdiction over the coops. In a long interrogation at this session, held on July 22, 1964, Senators Thurmond and Monroney, two senior members of the majority party (this was before Thurmond became a Republican) grilled me on what the FPC had in mind, and whether there was any point to extending regulation to the coops. I repeated several times the two reasons for the commission's decision: The growing role of the G&T coops in the interstate wholesale market, and the commission's effort to end uncertainty as to the area of coverage, which had already been applied to the investor-owned utilities. The committee members listened and received full answers to their questions. They had not called the hearing to be persuaded, though; they wanted to convince us of the folly of proceeding further.[6]

I left the commission before the issue was finally resolved. An appeals court held that Congress had not intended to cover coops within the Federal Power Act. Perhaps it relied on the interpretation provided by the Senate Appropriations Committee in the 1964 Appropriations Act, although the courts usually consider post-enactment interpretations by Congress to have little bearing on what Congress meant when the law was passed.[7]

I have described the coop jurisdiction episode at some length because it throws light on one of the ways that members of Congress can influence executive action, and, in this case, the way they can influence quasi-judicial action by an administrative agency. Congress always has the option to change the law, but this is the hard way for a member to work his will. The use of thinly veiled or implicit threats to hold up legislation or to reduce appropriations is usually enough. Bullying and harassment, at hearings or otherwise, are not rare weapons. Many, if not most, members of Congress in both houses are above such tactics. There are, however, other members who think it is demeaning to confess to constituents that they cannot influence executive or administrative actions.

The Role of Competition in Utility Ratemaking

Most administrators are aware of the desirability of developing ways to supplement the direct exercise of regulatory power. One effort made by the FPC during my tenure was to introduce competition by example. Each year we announced a list of generating plants that had the lowest heat rates—that is, which used the least amount of heat to generate a kilowatt-hour of electricity. This is a key test of a generating station's efficiency. We also publicized a list of the annual cost of electricity for consumers, so that they could see where their providers ranked. These and the other data were too gross a basis for determining which utilities should be rewarded or punished, or even which were doing the best management job. They did, however, put pressure on utility companies to avoid looking badly by comparison. We never claimed these data were factors in regulatory decisions, only pertinent information for consumers.

The FPC accepted the challenge to devise a workable and defensible system for measuring the quality of performance as a component of ratemaking. I found out that a group of midwestern utilities swapped copies of the comprehensive annual reports required by the FPC in order to derive a number of key ratios for financial performance and efficiency. The companies then used these ratios to determine where their strengths and weaknesses were, and what action was necessary for improvement.

The Northeast Power Failure

On the evening of November 9, 1965, I left the office between 6:00 and 7:00 p.m. When I got home twenty or thirty minutes later, my wife told me the president was trying to reach me, and that I should phone him immediately. When I phoned the White House number, I was immediately connected with President Johnson. Had I heard of the power blackout affecting all of New York City and much of the Northeast? I said I had not. (I did not listen to the radio broadcasts when I drove to and from work, and may have been one of the last people in Washington to learn of the power failure.) He said the situation was grave. New York City was immobilized. Lives were threatened. Looting had begun. There was danger of panic. The national defense communications system was affected, and its impairment could make the country vulnerable to attack (presumably by the U.S.S.R.). No one seemed to know who or what had caused the failure. He had asked the head of the Federal Emergency Management Administration, Buford Ellington, to find out what had gone wrong and what to do about it, but Ellington seemed to be at a loss on how to get started. Could I handle it? I said that I was sure I could with the FPC's resources. Johnson said that within the hour he would send me a directive to proceed, and that if I needed help in finding people or securing information, the FBI would be at my disposal. I phoned the commission's general manager, Harry Trainor, and the chief engineer, Stewart Brown, and asked them to contact the power system experts and supporting staffs, as well as the other commissioners. David Freeman was also alerted and asked to return to the office. When I arrived, I found several people already at work gathering information and, in a surprisingly short time, all the key personnel were present.[8]

I quickly outlined what I had in mind. Despite the language of the presidential directive assigning responsibility to me, I told my colleagues that I was treating the investigation as a commission matter, that I expected each commissioner to play a role, and that the report would be a joint effort. We decided, as a first step, to assemble the country's best system engineers at a meeting the following day.

I asked Stew Brown to make a quick telephone survey of the utilities involved, in order to assess the status of their systems, including any suggestion of sabotage. The results indicated there were no signs of explosion or sabotage, a conclusion I reported promptly to the White House and the media. Next, I parceled out responsibilities to the other commissioners, such as investigating the blackout's effect on transportation, on the media, and on the safety and welfare of New York City residents. I went home in the early morning hours, but Stew Brown and his staff were still working, trying to piece together the information that was coming in.

The commissioners and key staff met the next day with representatives of the major systems that were affected. Twenty-eight U.S. systems suffered interruptions for various periods, but we called in only those likely to be of greatest help. We also invited the Hydro-Electric Power Commission of Ontario. It was strongly interconnected with the U.S. systems and had also suffered a blackout. The only objection came from Charles Eble, the head of the Consolidated Edison Company [of New York]. I asked him to send a key system engineer, who was to bring with him the logs and other relevant records for the crucial period in which the blackout occurred. Eble demurred on the basis that the company would need the records for its own purposes. I said these records were essential if we were to find out what happened, and that I was authorized by the president to send in the FBI for the records, an action I then proposed to do. His hesitation vanished. In the pressures and excitement of those first few hours, I gave little thought (but I did give a little) to the question of the president's authority to take possession of Con Ed's records in this peremptory way.

The meeting brought out many of the key facts and suggested one or two hypotheses, but produced no clear answers. By this time, we were besieged by the press, radio, and television. Cables and other broadcasting equipment were strewn throughout the corridors. Nevertheless, this and succeeding meetings were closed to the public. Engineers are reticent enough in private about mentioning their guesses and hypotheses. In the public eye, they would be silent, and rightly so, because this was not a time to feed the media appetite for speculation and rumors.

Our ad hoc committee agreed that the series of events that created the blackout probably started in the area where the Ontario and Niagara Mohawk systems were tied together near Niagara Falls. We did not yet know, however, whether the origin was on one system or the other, nor could we identify how the initial events had cascaded to cover most of New England, much of New York State, including New York City, and a large part of Ontario. A total of 30 million people in an area of 80,000 square miles had been involved. One of the engineers pulled me aside and told me that, from what they had learned about the sequence of events, they believed the triggering event occurred in Ontario. It would have been foolhardy, though, to mention this without verification. If wrong, the engineers' assertion could have created resentment and criticism in Canada. Late in the afternoon, our meeting was adjourned until the following morning. The evening and the early morning hours were devoted to further collection and analysis of data.

In the morning we resumed the meeting. The additional data strongly supported the hypothesis that the sequence of events that caused the nation's worst power failure must have started in Ontario. I was reluctant, though, to rely on secondary evidence derived from analysis of the split-second timing of power flows. Fortunately, the mystery was resolved late that afternoon. Ontario Hydro had drawn the same conclusions we had. They made an intensive examination of their system and were able to pinpoint the triggering event. The point of origin was a relay at the Sir Adam Beck #2 Hydro Plant ("Beck") on the Niagara River in Ontario. A relay (a switch used to prevent overloading of transmission lines) started it all. The relay did not fail. On the contrary, it disconnected the transmission line that it controlled at the level of power flow for which it had been set. The setting of the relay (the maximum power flow that could take place before the relay automatically took the line out of service) was well below the capacity of the line. The setting was conservative, but not incorrect, when initially made. It had not been reset as loads grew.

At the time of the relay action, sixteen seconds after 5:16 p.m. (a peak load period), Ontario was importing from the U.S. about 1,500 MW (1,500,000 kilowatts) over five transmission lines. The loading was exceptionally heavy because of generation outages on the Ontario Hydro system. While 1,500 MW were flowing north, the New York State systems were transmitting some 1,800 MW to the southern and eastern United States, also a heavy loading. One of the five transmission lines that was carrying power north into Ontario went out of service because the load exceeded the relay setting. The power that it had been carrying was then automatically transferred to the other four lines, which overloaded them all. Within two-and-one-half seconds after the first line shed its load, all of the lines connecting the Ontario and New York systems, one by one, went out of service. The generators feeding the lines could not drop load instantly. As a result, the 1,500 MW that had been flowing north reversed course and were added to the power flowing south over the lines of the U.S. systems. These lines, in turn, could not handle the load. The Beck Canadian units automatically speeded up to respond to the need for power and were brought to a halt by safety equipment. Similar disengagements took place at other generating plants. In short order the entire New York system and some adjacent systems were caught in a cascading vortex.

As soon as the details become clear, I notified the White House and stepped into the corridor for an impromptu press conference. Reporters and camera crews had been waiting. I spoke only briefly, but provided the significant facts. Other radio and television programs were interrupted for this broadcast.[9]

What made the power failure especially serious was the hour at which it occurred and the difficulty and delay in restoring service, especially in New York City. The blackout began in late afternoon as most workers were leaving their offices for home. Subway systems, other electrified transportation, elevators, streetlights, and traffic signals were dead. All the buildings in the city, except for a few with their own generating plants, were dark. Gasoline could not be dispensed because the pumps were electrified. Appliances, television sets, lighting, and cooking equipment were all inoperative. Only battery-powered radios linked people to the outside world. Fortunately, broadcast radio service was soon restored. For most New York City residents, the blackout lasted until the next morning.

The blackout would not necessarily have attained historic dimensions but for a stunning error on the part of the system operator on duty in the Con Ed control room on that fateful evening. He had responsibility for coordinating with interconnected power systems and for protecting his own system. The system operator is in constant and friendly communication with his peers in other systems. When dollars are at stake (as in purchase and sale transactions), some bargaining may be involved, but system operators respond instantly to requests for help from systems in trouble when such assistance does not endanger one's own system. The system operator on duty had never before been presented with the stark choice of letting his own system collapse or denying help to others. Con Ed had enough capacity available after the initial power surges for its own needs, but the operator maintained his interconnections with other systems in trouble. Later, when some of his plants failed because of overloads, he could have saved the day by pushing a few buttons to disconnect enough of Con Ed's own loads to bring capacity and demands into balance. With a generating plant or two in operation, he would have had enough capacity to start the other plants, matching capacity and demand step by step. He did neither. Apparently, he panicked and froze in place until the last generating plant ground to a halt. The need to bring in power from outside sources to start the Con Ed machines greatly complicated and delayed restoration of service.

An example of corporate decency stands out in this dreary tale. Con Ed never revealed the name of the system operator or sought to place personal blame on him. After a period of leave, he was transferred quietly to another position that did not require him to make decisions under emergency conditions.

I have probably written more about the blackout than the subject deserves. Let me only add that it revealed generic flaws in power system planning, and the need for universal reliability standards, more explicit guidelines for system operations, better organization and equipment for control rooms, constant operator training, and prompt reporting of power interruptions to a central source. With FPC supervision, the regional reliability councils were established under the leadership of a national council to set and police standards. New FPC rules required immediate telephone notification to the FPC of any system interruption, as well as comprehensive reporting after the event. The report I made to President Johnson at his Texas ranch also emphasized the need—in hospitals and by other users where 99.9 percent reliability was not adequate—for standby equipment that would come on line as soon as central station service was interrupted. Most of the recommendations have been carried out, but not all. Gasoline pumps are still largely dependent on electricity, and I doubt that adequate provisions have been made to move people expeditiously out of most tall buildings or from stalled elevators when a blackout occurs.

CHAPTER 20

The National Power Survey

Another major initiative during my tenure was the National Power Survey (NPS). It was a comprehensive study of the status and problems of the electric power industry. It also contained recommendations for the industry's future growth and development. The results were published in 1964. The idea for such a study came from my TVA experience. I realized that a few power systems were working hard to incorporate new technological developments in their planning and construction, but most systems were laggards.[1]

In late 1961, with the endorsement and approval of my fellow commissioners, I sent to the White House a proposal that the FPC carry out a national power survey. The proposal was circulated by the White House to all agencies with a direct or potential interest, such as Interior, the Army Corps of Engineers, Agriculture, and several others. All of them endorsed the project except the secretary of the interior, Stewart Udall. He claimed that Interior was the proper agency to carry it out. Udall outranked me and might have hijacked the project, but either the president disagreed on the merits, or thought I had finder's rights because I had developed the proposal. At any rate, he approved my proposal, assigned the project to the FPC, and allocated the relatively small amount of money needed to get it started. I announced the project on January 23, 1962.[2]

The power staff and the commission met and tentatively designed a series of studies. The commission decided to set up a committee system made up of industry people. Some basic policies were adopted. First, the FPC should take full responsibility for the reports, looking to industry only for policy advice and as one source of data. We also requested that every committee should include one or more representatives from each of the three industry sectors—investor-owned, public, and cooperative. Since the public and cooperative systems together accounted for only a fifth of the industry in terms of investment or number of consumers served, this policy meant that they would be over-represented in relation to the private systems. This did not prove to be a problem, though. Representatives from the private systems were not worried or sensitive on this score, and, at any rate, the agenda excluded the matters in dispute between the industry sectors.[3]

The committees were to be advisory only; no votes would be taken. We decided not to attempt to resolve the issues that divided the various sectors, but to build on their joint interests. We were ever conscious of latent hostility and jealousies, and gave much thought to avoiding occasions for ruffling sensitivities. We would rely primarily on the most progressive and least combative people in the various branches of the industry, people who were recognized as leaders either in administration or in technical specialties. To assure that the members did not stray too far from their assignments, a member of the FPC staff would attend all committee meetings. Finally, the key committee would be an executive advisory committee (EAC), which would consist of chief executives and other industry leaders.

To get the project moving, I made early selection of an EAC chairman, Philip Sporn, the chief executive officer of the American Electric Power System, and a vice-chairman, G. O. Wessenauer ("Wess"), my former colleague and manager of power at the TVA. Sporn was probably the most knowledgeable and technically competent executive in the industry. He had the trust and admiration, if not always the affection, of the investor-owned companies. In TVA negotiations I had come to know him well. Strong-minded, innovative, plainspoken, work-driven, and tough, he was incapable of dissimulation. He did have a fault, though: He spoke too much at meetings and revealed too plainly his pride of opinion. Wess, too, was a technically competent engineer, an able administrator, and a skilled negotiator. Like Sporn, he walked a straight line and turned square corners. He had the respect and support of the public and cooperative sectors.

Another important committee was the general technical advisory committee, headed by Robert Brandt of the New England Electric System and Charles Almon of the TVA. This committee supervised and reviewed the work of the other committees, except the legal advisory committee. There were special committees on transmission and interconnection, future power requirements, generating stations, fuels and distribution, and five regional committees. The committees established subcommittees, as needed.

The EAC met eighteen times, always in the commission's meeting room. Sporn and Wess sat with the commission. The other members sat in a row of chairs facing the commission. The interested staff disposed themselves in chairs in the back or on the sides of the room. The colloquies were between the EAC members and the commission, not among the members of the EAC itself. Discussions were far ranging; only matters relating to the public/private power fight and contested legal issues were barred. Many of the questions did not relate directly to the NPS, but more broadly to some aspect of the industry: the prospects for growth; the effect of industrial migration; the adequacy of the industry's research efforts; the desirable mix of generating units; antitrust considerations; the adequacy of the transmission system and ways to strengthen it; the limits of the trend towards increasing scale in generation and transmission; the optimum amount and kinds of reserve capacity; standards of reliability; the extent of potential improvement in integration; and relative costs and benefits. Minutes were kept and distributed. Progress reports were standard fare, as well as discussions on the scope and direction of technical studies.

There was also much discussion of establishing national standards. The question was not merely technical. The adequacy and reliability of electric power supply depended on uniform observance of general standards. For example, if one of a group of interconnected companies decided to maintain a low margin of reserve generating capacity, or scant transmission capacity, it would impose operating risks on all the interconnected companies. Lesser problems might result in voltage or frequency fluctuations, which could injure generating or transmission equipment, or cause difficulties and damage to the equipment and operations of retail customers, particularly industrial ones. Uniform voltage levels are therefore important. If each company picked a voltage of its own, interconnections between systems would require large and expensive substations to change the voltage from the delivering system to that of the receiving system. Uniform standards reduce construction and operating costs, and provide a basis for the allocation of joint costs. The FPC had no jurisdiction to establish or enforce standards, and the problem was not solved by the time my term ran out.

NPS Results

The results of the NPS were twofold. First, the industry's rationalization was accelerated; second, the collaborative process itself conferred benefits. I had pointed out in some of my speeches to various groups that the United States was the only industrialized country in the world that had no plan to insure that the operation of the electric power industry benefited the entire nation. True, the American power industry was in many ways a model, but we had failed as a nation to take full advantage of the economies of scale in generation and transmission technology. Nor was there any systematic planning to make the best use of fuel resources and capital by siting plants near the fuel sources, rather than within the respective operating areas of the individual companies. We had only begun to tap the potential savings from inter-system and inter-regional planning, and from the coordination of construction and operations. While the overall efficiency and dependability of electric supply in the United States compared favorably with most other countries, this was because we were blessed with such great natural resources and the capability of a technologically advanced and productive society. Major benefits of scale and coordination fell into our laps because of the combination of the tremendous size and rapid growth of the American economy. The time had come, however, to plan for optimization of benefits, pursuant to Section 202(a) of the Federal Power Act (as amended in 1935). In short, the commission needed to encourage the voluntary interconnection and coordination of the nation's electric power facilities. It was apparent that future savings in costs were to come from joint planning, and from the construction and operation of adjoining systems on an integrated basis. I kept pointing out that this was possible simply by working together. It did not imply loss of local identity or autonomy of the individual companies.[4]

The NPS idea was not particularly unique, but it was timely, well executed, and fruitful. It was as though the power industry, struggling for a sense of identity and purpose, found in the NPS a medium to establish necessary standards, coordination among companies, integration among regions, better use of new technologies, and a new dedication to reducing costs and improving service. The industry people who worked on the NPS believed the project was important, and, in so thinking, made it so. One reads occasionally in the trade press that the early 1960s were the finest period of the electric power industry. The NPS helped make it so.

One early surprise was to find out how well the public power and investor-owned people got along with each other. Like the Hatfields and the McCoys, the two sides normally needed no particular reason for feuding. At NPS meetings, however, they were on their best behavior—all smiles and given to polite and friendly exchanges. More important, they worked together and productively for the first time. It was in the FPC meeting room that each side met its peers. An example was the friendship that developed between Wessenauer of TVA and J. Harris Ward of Commonwealth Edison Company [of Chicago], which resulted in the construction of an interconnection between the two systems.

The stimulation of energy research and the creation of the Electric Power Research Institute are two of the most dramatic and important illustrations of the serendipitous effects of the NPS process. I had often complained at EAC meetings about the low level of industry research. At one time, the electric equipment manufacturers, especially General Electric and Westinghouse, had carried out extensive research programs, including a modicum of fundamental research. (I use "research" in a broad sense to include development and demonstration.) Then came the antitrust prosecutions of the 1950s, in which the manufacturers admitted collusive pricing of heavy equipment, paid large fines, saw some of their executives (briefly) put behind bars, and agreed to consent decrees forswearing these practices. When these companies gave up their lives of crime, they suffered severe attacks of pernicious frugality, which meant drastic cuts in research. There may even have been a bit of desire to demonstrate that the government had erred in upsetting their cozy arrangement. The new policy was to avoid research that did not have strong promise of leading to early and profitable commercialization. When the two largest manufacturers cut back, no one picked up the energy research burden. How, I kept asking the members of the EAC, could the power industry meet the research challenge?

The members of the EAC were sympathetic, but offered no solutions. Our discussions on research were becoming repetitive and probably boring. Finally, to cut off profitless discussions, Philip Sporn asked me why, since I felt so strongly about the need for more energy research, did I not make my case to the Edison Electric Institute (EEI), which represented all the investor-owned companies. I answered lamely that I had not been asked. Most members of EEI looked askance at me because of my public power background. I had previously been invited to speak to only one group of executives of the investor-owned companies. Sporn replied that securing an invitation was easy. All the other company representatives merely nodded. They had apparently agreed in advance with Sporn on this tactic. I told Sporn to go ahead, and a few days later I received an invitation to address the EEI's annual convention in Denver on June 4, 1963.[5]

I worked particularly hard on my EEI talk. It was a favorable opportunity to reach the industry, and the subject was close to my heart. It turned out to be about as good a set piece as I have ever managed. I called it "The Special Research Problems and Opportunities of the Electric Power Industry," not a title that would grab the man on the street, but reasonably descriptive. After commenting on the unlikelihood that I should be addressing the EEI after spending most of my working life as a TVA employee, I added that the invitation had been offered in full knowledge of my background. I had accepted it "in the same spirit of adventure in which it was tendered." This eased the tension a bit. The research question, I said, had proved to be a recurrent theme at the meetings of the EAC. It was interesting and provocative because of the need for improved research, and the awareness that the structure of the industry made it difficult to secure that improvement. Technological pioneering had been the lifeblood of the industry since it was launched eighty years before. The industry's high growth rate, the immense investments (then over $140 billion), and the much larger investments that the industry would have to make in the coming decades made it imperative for the industry to learn as much as possible about the performance of the equipment and the materials from which it was made.

I pointed out, too, that the present level of top management had entered the industry during a period of rapid innovation in electric power technology. By the early 1960s, however, the brightest graduates of the engineering schools were opting for electronics, space, missile, and nuclear work. Scientific breakthroughs in the energy area, which once had come unbidden, now occurred only as a result of intensive and well-funded research. The electric power industry was still the greatest growth area of all time, but it could only assert technological leadership and attract talented people if it could hold out the promise of participation in the work of scientific discovery.

In preparing my talk, I had concluded after much thought that the best approach was the creation of an electric industry research institute. It would need to embrace all aspects of the industry if it was to succeed in winning wide support and generous financing. The standard for determining its program would be the needs of the industry itself, not the manufacturers or producers of raw or intermediate materials. No single representative from one of the sectors, whether public, investor-owned, or cooperative, could propose such an organization without meeting skepticism. That was why, although I was preaching to the converted in the meetings of the EAC, none had volunteered to address the heathen. An honest broker was needed. I thought that most industry leaders on all sides had come to trust me, some no doubt with their fingers crossed. I determined, therefore, not to rest with a scolding or lamentation, or even with a general pitch for more research or for an industry research facility. I decided to end my talk with a specific proposal:

> I have no prescription for curing the research problem of the industry, but I should like to propose a small step which could lead to some improvement. As the agency with over-all national responsibility in the power supply field the FPC could appropriately call together a group of knowledgeable people representing the private, public and cooperative sectors of the industry, the manufacturers, and the research community, to consider how the industry—acting as a whole—could best serve its own interests and the interests of the nation by stimulating more intensive research efforts in areas where such intensification is likely to be rewarding. Such a group would be fully aware of the risks of over-centralization and of the need, with respect to product research, for integrating the research effort with development and production.
>
> Specifically, such a committee would be asked to make recommendations with respect to the establishment of a permanent industry-wide organization, including its composition, responsibilities and administrative structure.
>
> The FPC seeks no role in such an organization. Our only interest is that the industry itself—all of the elements of the industry—shall create and maintain an organization which will assure the utmost technological advancement which can be brought about by research and development

> activities on a scale commensurate with the opportunities and promise of rewards. Here, indeed, is a goal on which all of the forces of the industry can unite.

I had expected that the people at the EEI meeting would debate my proposal over a period of time, or refer it to a committee for study. To my surprise and pleasure, the EEI passed a resolution approving my proposal and requesting me to proceed with appointment of an organizing committee.

In due course, the committee recommended the creation of a transitional agency, which was soon established. Dubbed the Electric Research Council (ERC), its role was to take over the small research program being administered by EEI, to develop some sample projects for industry-wide support, and to lay the foundation for a permanent industry-wide research agency. The ERC moved somewhat slowly, but in a half-dozen years it recommended turning over its agenda and workload to a fully funded successor, the present Electric Power Research Institute.[6]

A final by-product of the National Power Survey, though unrelated to my EEI talk, was the creation of the National Electric Reliability Council (NERC), which soon changed its name to the North American Electric Reliability Council. This change reflected the fact that the major Canadian systems had become members, though the organization retained its initial abbreviation, NERC. I believe the NERC has been a solid success. In setting reliability standards at a high level, and especially in prescribing the margins of reserve required to be maintained at all times for both generating and transmission capacity, it has assured a high degree of reliability of electric service throughout this country and Canada.[7]

The Fate of the NPS

In the last House hearing on FPC appropriations at which I testified (March 29, 1965), I spoke at some length about the NPS. On that occasion, I pointed out the importance of follow-up:

> I have already referred to the importance of the National Power Survey which was published by the commission on December 13, 1964.
>
> The commission's survey has encouraged the industry to increase its research efforts and as a result on March 15, 1965, the industry announced the formation of a research committee including representatives of all segments of the industry to accelerate its research efforts on a coordinated basis.
>
> The president has recognized the survey as a constructive step in directing the attention of the government and of all of the various interested industry groups to our greatly increased power needs of the future. The goal of the national power survey is lower-cost electricity for all Americans. To make this goal a reality will require a follow-up on the beginning which the

> survey has achieved in planning the coordinated growth of the electric power industry. The extent to which the ideas and suggestions in the survey report are adopted will depend in part on the degree of the commission's efforts in encouraging the industry to continue to work together and make the detailed evaluations prerequisite to implementing these suggestions in the survey report.
>
> The working relations which have been established between the commission and the industry and among the various segments of the industry are the most valuable products of the survey. They should not be allowed to lapse.
>
> Therefore, we expect to continue our power survey advisory committees, perhaps in altered form, so that the commission can continue to carry out its statutory duty to encourage interconnections in cooperation with all segments of the industry.[8]

My departure and Stew Brown's death from cancer at about the same time left the NPS project without the support it needed. My successor, Lee White, continued the NPS committees, and so did his successor, John Nassikas, a Nixon appointee, but with slowing momentum. A four-volume report, the *1970 National Power Survey,* was issued in late 1971. It was the last of the NPS series. The FPC (and FERC) abandoned the role of industry leader and gadfly, reverting instead to preoccupation with its regulatory role.[9]

Why was the NPS permitted to fade away? For one thing, the times were not favorable. The 1970s were a period of rapid inflation and ever-increasing costs. The attention of industry executives was focused on serious internal problems: maintaining construction schedules, raising money in a lender's market, and securing the rate increases necessary to maintain credit. On the commission's side, the pressure of the gas agenda could absorb all the commission's energies unless it budgeted its time so that it could also deal with the problems of the power industry, which were equally serious, but were not equally reflected in the regulatory agenda. It is only fair to add that when I left at the end of 1965, there was no one left on the commission with a background in the electric utility industry, or a concern for working out its problems.

The fate of the NPS points to a failure in the process of appointing commissioners, not only to the FPC/FERC, but also to other major regulatory agencies. To take the FPC/FERC as an example, most appointees are not required to have had any experience in operating, managing, or even regulating the two vital industries subject to regulation. A general legal, accounting, or economics degree is considered fully adequate. Sometimes the only obvious credentials are U.S. citizenship and the absence of a criminal record, plus political acceptability and, usually, political sponsorship. Since FERC is a bipartisan agency, it is necessary to appoint two members of the party other than that of the president who are not seriously objectionable to the White House. The

Democrat presidents usually appoint liberal Republicans, and the Republicans find Democrats who think, talk, and act much like Republicans. The vice is not in the absence of qualified candidates who could be confirmed, but in the failure to seek them out. Regulatory appointments too often become pawns in the game of high-stakes politics. An influential senator of either party can bargain for a right of appointment to almost any regulatory agency (perhaps not the Federal Reserve Board), as long as the White House approves the nominee's basic qualifications and political views.

CHAPTER 21

The Natural Gas Program

The commission's powers over the interstate natural gas industry were very nearly plenary. At the time of my service, the gas pipeline companies [also known as transmission companies] combined the merchant and transportation roles, that is, they bought and owned the gas that they transported and sold, purchasing for the most part from gas producers, but also buying from other pipelines when such purchases were advantageous. When these companies were involved in interstate commerce, the FPC determined the prices at which they bought, as well as their resale [wholesale] prices. Thus, the FPC controlled the amount of both gross and net revenues of the pipeline companies.

The FPC's grasp on producers [of natural gas] was lighter. The Natural Gas Act specified that the FPC should have no jurisdiction over producers, but the Supreme Court eventually held that the FPC could regulate the sales of any independent producer that sold its gas to an interstate pipeline company. Despite possessing this authority, the FPC had made no real headway in determining how to fix the field price of gas. Moreover, because all the relevant producer rate cases were up in the air, it was impossible to determine the revenue entitlements of the pipelines. The situation was aggravated by the "pancaking" problem—the filing of second or third rate-increase applications, which went into effect subject to refund, before the first rate application was decided.[1]

Everyone was furious at this stalemate. Consumers were being required to pay increased rates without the benefit of a ruling affirming that these prices were fair and reasonable, and without knowing when, if ever, they would receive refunds. The payments by pipelines to producers were subject to refund, relatively small amounts for the giant oil companies, but serious burdens for some of the small producers. Most of the latter had so much at stake in potential refunds that the outcomes would determine if they were operating in the black or in the red. Their stockholders and bankers shared in the problems created by the uncertainty and confusion.[2]

Cleaning Up the Backlog of Pipeline Rate Cases

Our decision was to hold brief hearings to determine fair and reasonable interim rates to be effective until the Permian case was decided, and to make these rates final and not subject to refund. We used published sources for the data, and made the best guesses we could under the circumstances. The courts were kind, recognizing that there was no better solution. Even with this strategy, the pancaking problem would not have been resolved at any early date without another indispensable tool, a fast settlement program. In a few months, FPC staff developed estimates of what each case (or a consolidated case covering all the requested increases for a given pipeline) might be worth after a full trial. Settlement conferences were called in which all the parties participated—the pipelines, the distributors, the consumer groups, the municipal and industrial intervenors, and any others who had a stake in the result. Since all parties benefited from a fair settlement, they usually agreed on figures close to the staff's recommendations. The results were promptly brought to the commission and fully explained. The settlements were at levels much lower than the suspended amounts, and they released a tremendous amount of money as consumer refunds. At the same time, they enabled producers and pipelines to retain a fair share, and to give them solid and bankable earnings figures.[3]

Not everyone was pleased with the settlement program. The lawyers, lobbyists, and professional witnesses, who had confidently counted upon putting their grandchildren through college on the fees they would receive for their work on these interminable cases, now had to scratch for business. The FPC assistant general counsel in charge of hearings protested strongly. Settlements were fine, he said, but they should be entered into only after completion of formal hearings—this despite the fact that it might be many years, and possibly not until the third millennium, before these cases were concluded—and before consumers (or their grandchildren or more remote descendants) would finally receive the refunds to which they were entitled. When I declined to halt or change the settlement program, he resigned, though not at my request; he was a good lawyer, but could not adjust to a different way of handling cases. Most of the staff were exhilarated by the prospect of performing in a way that served the public interest and would offer them far better experience. They were also pleased by the degree of trust and harmony that had developed between them and the commissioners.[4]

The Expedited Decision Procedure

The Natural Gas Act provided that certification for new construction could be granted only after a hearing. Many cases were routine, but they were assigned for a hearing before an ALJ. Even when briefs were waived, they proceeded slowly to the commission. We changed all that. We defined a category of certificate cases that involved less in dollar investments than a prescribed maximum, and that did not raise any significant policy questions. In such cases, the pipeline's interest in system improvement and expansion coincided with the public interest in better and more widely available service. These cases came before the commission on a staff recommendation. The commission's consideration of the recommendation was treated as a hearing, even though less than a minute might be devoted to it. This procedure broke the certificate logjam. Now the commission could devote itself to cases involving large amounts of money, gas, or important issues.[5]

As best I recall, there was only a single case in which the expedited decision procedure was abused. One of the largest pipeline companies, Tennessee Gas Transmission (TGT), used the procedure to attach a large gas reservoir in the Gulf of Mexico to one of its pipelines, on terms that violated FPC policies on rate ceilings. This was no routine case. Only a couple of miles of line were needed to attach the reservoir to an existing TGT pipeline, but hundreds of millions of dollars and vast amounts of gas were at stake. The pipeline management may have thought that they had achieved a fait accompli, but they were mistaken. We reopened the case, and my comments at a public hearing were tart. The FPC took prompt action. The commission's action was appealed to the courts, where it was upheld.[6]

Interactions with the FPC

The major jurisdictional utilities are frequent players before the commission. They come to know the commissioners well from their opinions, speeches, and, in recent years (after my departure), from attendance at commission meetings under the sunshine laws. Similarly, the commissioners learn a lot about the companies and their managements. For this reason, it is in the company's interest to avoid giving cause for complaint due to disingenuousness in dealing with commissioners and staff. In my time, and I believe still, the ethical standards of the practitioners before the FPC were extraordinarily high.

The same high standards were not necessarily displayed by the lobbyists or some executives. The worst example of excess related to the so-called "ten-pound questionnaire," which provided the factual basis for the area rate cases. Since the FPC did not regulate the producers per se, there were no filings or reports from which we could derive necessary information on the costs of drilling and completing wells, costs of gas and oil separation, or the myriad other facts required if the proceedings were to be based on more than guesswork. To fill in the gaps, the FPC prepared and released a comprehensive questionnaire. It was necessary to include instructions that would produce realistic and comparable figures from the producers. A questionnaire, including instructions, weighed about a pound. A separate form was required for each of ten areas, but only the largest firms operated in all these areas, and ten copies were sent to each of them. Small producers were asked to fill out a simplified form, though their participation was voluntary. As soon as the forms were mailed out, a lobbyist showed up on a major television news program with the entire stack to provide the comedy note to which producers of commercial news broadcasts are addicted. The commentary falsely said, or implied, that all producers were required to complete the whole stack. It made good theater as an example of rampant bureaucracy, and enjoyed wide newspaper and magazine publicity.

I had no advance notice of the broadcast, and, of course, no opportunity to appear. I asked the television network for a chance to respond, but was rebuffed. I did explain the facts at a press conference, although the explanation never caught up with the story. Many representatives took up the cause of the producers, especially those from the oil patch. It was also raised at congressional hearings, but here it was possible to explain why determination of fair production prices hinged on obtaining answers to the questions listed in the questionnaire. I must add that I thought the network's failure to check with me about the accuracy of the lobbyist's story, not to mention its refusal to give me a few moments to explain the need for the questionnaire, was a betrayal of its responsibilities as a channel of public information. How can a lobbyist be trusted to tell a straight story? Anything for a good laugh about bureaucrats; to hell with explanations that provide solid information and expose the hollowness of the joke. Many years later, it still rankles.

Opening Avenues of Communication

The FPC carried on its work during my tenure with open avenues of communication with the regulated industries, particularly natural gas. I have already mentioned the adoption of a policy of inviting key staff members to all commission meetings except those dealing with personnel problems or reports on delicate legislative matters. We required no secrecy oaths, but did expect discretion in passing along indiscreet statements by commissioners. We also cautioned against premature disclosure of decisions that might have a market impact. There were some leaks of distorted versions of what I had said by Howard Morgan; otherwise, we had no reason for complaint. I might add that I never authorized or conducted an investigation of alleged leaks because I thought they would be unprofitable, if not counterproductive. Our staff trusted us, and we trusted them, which is the best kind of security system. Staff participants, however, were free, and sometimes encouraged, to pass along to their subordinates information on how we worked and the substance of our deliberations. I assume this information soon permeated the organization and that much of it became available to the practitioners. The attendees could certify, at the least, that there were no shenanigans, unless the votes of the political appointees on matters of interest to their senatorial sponsors may be said to come under this heading.

The most active of the gas industry executives was William Elmer, the chief executive of the Texas Gas Transmission Corporation (known in the trade as Texas Gas). He was an early well-wisher. After I joined the FPC, he came to see me several times, usually with good or plausible suggestions for general changes in our rules. At one of these meetings, after the NPS was launched and its committee structure was established, he suggested the need for an executive advisory committee for the natural gas industry. I was not immediately taken with this proposal. On the electric side, the committee's dominant role was to facilitate preparation of the NPS. No similar survey was contemplated on the gas side. But the electric industry committee had developed a life of its own and demonstrated its usefulness as a communications medium. This advantage was vigorously argued by Elmer, who also made the point that to create such a committee for only one of the two industries we regulated smacked of favoritism in the constant struggle between gas and electricity for larger shares of the energy market. The latter point registered, and the commission approved creating a Gas Industry Advisory Committee. Unlike its electric counterpart, there was no substructure of industry committees and subcommittees.[7]

The Gas Industry Advisory Committee was perhaps the most doubtful communication medium, although it proved useful. It was doubtful because the line between proper and improper subjects was hard to draw; it was useful because the commissioners were no longer "faceless bureaucrats" to the executives of the industry, but individuals they came to know at close quarters. I am confident that as the committee members came to measure each of us, they became less suspicious of motives and more willing to give us the benefit of the doubt. The relationship did not reduce the amount of litigation; however, it did bring down the level of suspicion and recrimination, especially among the pipelines. It also provided each group with more confidence in the bona fides of the other. These are not inconsiderable benefits.

Did we go too far in consorting with the officials of the regulated companies? One could make a case that we overstepped the bounds of propriety. My colleagues and I demonstrated in our regulatory actions that we were severe, but not punitive. We illustrated that we were motivated by the public interest, that we did not hesitate to set rates at levels far lower than those for which the industry fought, and that we were willing to order hundreds of millions of dollars in refunds. The consumer groups were happy with the conduct of the organization and with the results, and our reputations for probity and protection of consumers were untarnished. While our predecessors avoided any open engagements with industry people, lobbyists roamed the corridors. Their employers must have thought they earned their keep. Another difference was that our predecessors held their meetings in strict privacy. Consumers were rightly suspicious.

I do not mean to suggest that these opposites are the only choices. There may be a middle way, but middle ways are apt to require complex regulations and an enforcer. The hampering effect on supervision by an officious and virtually independent subordinate involves tremendous but unquantifiable costs in conducting the nation's business. Is a snoop in every department and agency, aided by a staff of assistant snoops, an advance in public administration? I doubt it. The choice is not between a central snoop system and rampant crime. A well-run agency will police itself.

I distinguish between snoops and whistle blowers. The latter do not ordinarily snoop for a living, but expose suspect transactions that come to their attention in the course of their work. Naturally, they should report fraud or abuses. At the same time, an environment of distrust can breed overzealous whistle blowers, including people with their own scores to settle. And, if whistle blowers are installed to spy on an agency, as sometimes happens, whistles will sound at all hours.

CHAPTER 22

Congressional Conflicts: Albert Thomas and Other Representatives

The *bête noire* of my tenure, Howard Morgan not excepted, was Representative Albert Thomas. He represented the Houston area, then and still the petroleum capital of the world. Thomas therefore had more than a passing interest in the decisions of the FPC. The attention he gave to me became a source of embarrassment and painful concern. Most representatives, including most of those from Texas, left me alone or made only perfunctory contacts as a part of their constituent service activities. Not Albert Thomas. His attentions were assiduous and forceful. Of course, Al Thomas was not just any representative. He was a senior member of the House Appropriations Committee and chairman of both the Independent Offices Subcommittee and the Deficiencies Subcommittee.[1]

At hearings of the Independent Offices Subcommittee, Thomas was frequently the only member present. Other representatives might attend for a while and ask some questions to record proof of their attendance, but rarely did they dissent or express a view inconsistent with that of the chairman, unless it was as a matter of upholding minority party doctrine or positions. For practical purposes, Thomas was the committee, and the fate of an agency could well depend on maintaining good relations with him. Naturally, if the agency head expected personal favors, such as an assigned car and driver, or an allowance to the agency for the entertainment of foreign visitors (called a "representation allowance"), the individual had to jump when Thomas gave the signal.

In his capacity as chairman of the Deficiency Subcommittee, Thomas may have had even greater leverage. Indeed, with deficiency appropriations, as in the case of the regular full-year appropriations, the requests are made by the president after informal hearings before the White House Office of Management and Budget. Before the end of a fiscal year, many agencies need to be rescued from poverty or insolvency, either because the initial appropriations were too low, or because of unanticipated outlays. The alternative to a deficiency appropriation may be a curtailment of service or some payless paydays. The White House is necessarily involved in these crises as it seeks to defend its programs and personnel. In such cases, the leverage of the chairman of the Deficiencies Subcommittee is at its most potent.

Thomas was ramrod straight, punctiliously dressed in an old-fashioned way, and courtly in his manners unless he ran into an obstacle. At hearings, he was always well prepared. He did his homework, and his questioning was sharp and pertinent, though it sometimes amounted only to niggling over details. If he asked for a number, he wanted a prompt answer. I was always the witness for the FPC, but as customary in budget hearings, I was accompanied by the other commissioners, the general manager, key department heads, and the FPC budget experts. The substantive questions on programs, policies, objectives, progress, and the basic numbers I could handle without help. I did not, however, keep all the details in my mind, and neither did the budget experts. Some of the questions verged on efforts at entrapment and required a frantic search by the budget people through large notebooks. If the answer could not be dredged up instantly, Thomas turned impatient and sarcastic. I understood he treated other agencies much the same way. He was probably more stern with the FPC than with agencies with which he had no serious policy differences—or perhaps those agencies were more malleable. Since the hearings were private and Thomas was free to "correct" the transcript (which included striking or changing anything he said before the hearings were printed), his indiscretions could be buried. Our appropriations suffered in his subcommittee, but the full Senate Appropriations Committee and the Conference Committee usually came to our rescue.

What did Albert Thomas really want? He wanted to control FPC's policies on the pricing of natural gas in the field—that is, the price the pipelines paid to the producers, which the pipelines passed along to the gas distributors, who, in turn, passed them along to the retail gas customers throughout the country. Soon after I took office, Thomas invited me to have a cup of morning coffee with him in his office. The invitation was a command and I accepted promptly. Similar invitations were extended from time to time thereafter, until Thomas decided he was wasting good coffee. These morning coffees with Thomas became a great strain on me. They followed a pattern. The question was always the price of gas, either in general proceedings or in connection with applications by particular companies to buy or sell gas. Each time I would explain our procedures for determining gas prices and the need to avoid ex parte discussions, but he paid little attention. He pressed again and again. I explained several times that under FPC procedures I could not create exceptions. The other commissioners and the staff were vigilant, and they would quickly catch me in any inconsistency. Thomas said if I wanted to oblige him, I could find a way. He insisted that I dominated the commission and could do as I wished, a view that was totally wrong, as I vainly tried to explain.

At one of the last of our coffee meetings, Thomas (I always called him "Mr. Chairman") played what he may have thought was a trump card. A small producer, who happened to be a Jew, wanted to sell a relatively small amount of gas at a higher than authorized price. Thomas must have thought I could not turn down a coreligionist who wanted to sell only a "bucketful" of gas. I answered in the same way as in the other cases. Thomas was obviously disgusted with what he apparently considered a breach of my religious obligations. I was saddened that so worldly a man should carry in his head such a stereotype of the Jew.

Thomas finally enlisted the White House in his efforts to influence me. I was surprised one day when Walter Heller, the chairman of the Council of Economic Advisers, a distinguished economist on leave from the University of Minnesota, asked to see me. Although we had met several times at the White House mess, we were scarcely intimate. It turned out that his mission was to tell me that I was creating problems for the administration by antagonizing Thomas. I should, he said, take a softer line. I could hardly believe I had heard him correctly. That Heller would serve as a messenger to deliver what amounted to an ex parte communication on gas prices was hard to credit. I told him that I had done everything I could to appease Thomas. If deference and efforts at friendliness were enough to satisfy him, he would have no reason to complain. Heller did not seem to be satisfied with my response, but I could do no better. I sympathized with Heller since he would now be required to report to [Larry] O'Brien that he had failed in his mission, a mission he could not have relished. Why Heller, who was not in a position that carried political responsibilities, was picked, and accepted, for this unsavory duty was not clear to me.[2]

A little later, O'Brien asked me to visit him at the White House. In a businesslike way, he showed me a paper he said Thomas had given him. The details of the paper I have forgotten. It consisted of an incoherent charge of some unfairness on my part or on the part of the FPC to a producer. There was not enough information in the paper to enable me to determine the basis for the complaint, and I could only say that I did not know about the grievance or its basis, though I would seek information from the staff. I did so when I returned to the office. The gas people said it might refer to a transaction with which they were familiar, but, if so, the paper was in error. I examined all the relevant material and could find no good basis for complaint. I paid another unsatisfactory visit to O'Brien, who had nothing he could add to clarify the picture. The occasion passed, not without leaving Thomas with a residue of resentment, I am sure.

In the course of our talk, O'Brien said Thomas was threatening to refuse to recognize me as the commission's chairman for purposes of his hearings. I said he could not bar me—I had done nothing to offend him, and his only grievance was my refusal to accommodate him in carrying out my responsibilities. I assume O'Brien passed along my response. At any rate, the threat was heard no more.

I was not asked, either by Heller or O'Brien, to do anything improper. Rather, their effort took the form of seeking assurance that I would not be unduly severe or discriminatory in producer matters. If I knew Thomas, such assurances would not be enough to satisfy him. To say the least, the approaches of Heller and O'Brien were unwise. For my part, I was filled with regret that I was a cause of embarrassment to O'Brien in his relations with Thomas. Kennedy was having a difficult time with Congress in attempting to move along a comprehensive legislative agenda, including basic civil rights reforms, and he needed all the help he could get.

When I paid a visit to Thomas at his office, I would frequently see a group of oil industry lawyers and lobbyists waiting in another room. They seemed unembarrassed, relaxed, and in no hurry. I was ushered in to see Thomas while they waited. Perhaps they had come for a report from Thomas on his talk with me. Their presence reflected the closeness of their ties with Thomas. Of course, he was not the only member of Congress with whom they had close relations, but he was one of the most influential. Vice-President Lyndon Johnson was also the object of their attentions, and it may well be that O'Brien was acting on suggestions from Johnson when he made contact with me.

Thomas had two ways to avenge himself, in addition to attempts to create problems for me at the White House: (1) abuse me at the hearings he conducted; (2) deny needed appropriations. As to the former, his manner was frequently acerbic, and many of his questions dealt with petty details. Typical queries asked for a breakdown, by grade, of existing groups of employees or those we proposed to hire. Normally, these questions were handled on a staff level, between his people and ours. No one else in my long experience of testifying before budget and appropriations officials had inquired about such matters. When Thomas was given the answer, there was never a follow-up question or comment. Testifying was part of my job, and he was entitled to ask the questions. I was under an obligation to answer as best I could, but I thought he was bearing down pretty hard on trivia. I was rather proud of my ability to present to the committee a coherent, factual, and persuasive defense of our appropriation request. I did not appreciate his putdowns.

In one of the hearings on Independent Offices Appropriations for 1965, Thomas charged at me in the first few moments. Referring to our request for some additional positions, Thomas said, "We have seen a lot of empire builders, but we give you the No. 1 spot." I replied meekly that this was an honor I had not sought. I went on to explain that our predecessors had left us with monumental backlogs to clear up. Thomas treated this as a personal affront. "[W]e are not as smart as you are, but we are not the dumbest folks in the world. Do you mean to tell us we did not know what we were doing here for ten or twelve years before you came along and that you are the Messiah who is going to save the country and show us where we are all wrong? Your egotism is most astounding." Representative Evins of Tennessee, a member of the subcommittee, tried to come to my rescue when I referred to the *Landis Report*'s conclusion that the FPC was the worst of the federal regulatory agencies. Thomas reacted with disbelief. Evins replied: "At the present time we commend the former chairman and former commissioners for doing a good job, but as I recall at that time, they were given hard going because it was said they were not doing their job as they should. . . . At the time Commissioner Swidler came in as chairman of this commission, for the very first time there was introduced a new look and a new, fresh approach. I think he should be given a proper opportunity to carry out his enforcement authorities as he interprets his duties as chairman of the commission." This statement helped, but did not prevent more hostile personal criticism at the hearing. I held my temper throughout and responded factually to Thomas's questions and charges. It was a punishing ordeal.[3]

My patience was due in part to my knowledge that responses in kind would produce a crisis harmful to me and perhaps to the administration. I knew that I was being provoked deliberately in the hope that I would commit an indiscretion. In some small measure, my restraint was due to my recognition of Thomas's courage and my admiration for his work habits. I had learned that he had cancer of the spine, from which he died not long after I left the FPC. Yet he sat straight in his chair and asked for no favors or help from committee members or staff. He was always familiar with the details of appropriations requests, rising early to study these dreary documents before coming to his office. It was a pity that what he wanted so badly from me I could not give him. In many ways he was a model member of Congress, though he could not serve two masters.

There were only two other improper approaches from representatives that I can recall. Several let me know in one way or another of their interest in particular cases, but they stayed within permissible bounds. Representative Charles Buckley, chairman of the House Public Works Committee, called me to his office to say he hoped the commission would approve Con Edison's pending application to construct a pumped storage plant on the Hudson River at Storm King Mountain (not far from West Point). His interest stemmed from the fact that he owned land near the site whose value would greatly appreciate if the plant were built. I told him it would be improper for me to hear any more, and left his office.[4]

Another instance of intervention by an individual member of Congress occurred at a party given at his house by a friend of TVA days, Edward Falck. One of the guests was Senator Russell Long of Louisiana, who was chairman of the Senate Finance Committee. He had imbibed at least one drink too many when he grabbed both my lapels, perhaps for support, to tell me that he had large oil and gas interests and wanted a high gas price in the pending Louisiana area rate proceedings. His condition made it possible for me to detach myself gracefully.[5]

There was congressional interference in other ways. One New York representative, Richard Ottinger—who exerted even more pressure on the FPC than the bad guys—was forever writing and telephoning me in the interest of public power groups. The same was true of a representative from an area of California with several municipal electric systems. These individuals complicated my life, and their persistent attentions included suggestions of congressional investigations. They considered their pressures were for a noble purpose, but they cast an additional obstacle in my efforts to decide cases fairly.[6]

The ability to "deal" skillfully, to buy support, to maintain administration priorities, and to avoid setting harmful precedents or legislative rivalries are among the greatest political gifts. In another life, I should like to try the political experience, but that is not the road I have traveled. My own work has provided little room for dealing and bargaining for political advantages.[7]

CHAPTER 23

Official Trips to the U.S.S.R. and London

The U.S.S.R. Trip

In 1962, the United States and the Soviet Union agreed to exchange delegations of visiting electric power experts. At that time, no official technical exchanges had taken place. The Soviet Union, with its vast distances and remote power sources in Siberia, had already solved the technical problems of 500-kilovolt transmission. Study of their 500-kilovolt structures could save a great deal of time and money as compared with reinventing what the Soviets had already achieved. Stewart Udall, U.S. secretary of the interior, was designated as the team leader. Each of the administrators of the agencies with electric power responsibilities was invited to participate, or to designate a substitute. The Interior delegation included Udall and James Carr. Charles Luce, head of the Bonneville Power Administration and, I believe, a representative of the Bureau of Reclamation went as well. Chairman Aubrey Wagner and G. O. Wessenauer deferred to Fred Chambers, the TVA's chief systems engineer. Lee White was another member. I was also invited. Udall told me that I was second in rank; this meant I would be in charge if he were absent.[1]

We spent the first couple of nights in Moscow. We stayed at the Hotel National, opposite the Kremlin. The hotel was modest by U.S. standards. It was a little musty, but not uncomfortable. A few stories in height, it had an elevator that could accommodate only the operator and two adult passengers. It moved very slowly. Most of the American group used the stairs to get to our rooms. Here we became familiar with the fanatical security concerns of the Soviets. On each floor, night and day, a woman (always a woman) sat at a strategically placed desk where she could command both stairs and elevators. She dispensed the room keys and required that they be returned to her when we left the floor. She kept a record of our comings and goings, and of all visitors. If anyone had learned how to put through a telephone call—there were no phone books or directions, and the hotel operator spoke only Russian—I assume that it would have been monitored. We were to see many other indications of the Soviet spy phobia, including billboards and street posters warning against spies and public discussions. We were accompanied everywhere by two Soviet aides who, as it turned out, were KGB agents (we had assumed as much from the beginning). The senior of the two was a man named Poliakov, if memory serves, who was about forty, blond, and of typical Russian build (medium height, sturdy, and solid). The other was a woman about thirty whose name I do not remember. She was rather pretty and always serious.[2]

Early in the trip, we had a head-to-head confrontation at a meeting with the Minister of Energy on the subject of taking photographs. The Soviets permitted us to bring in cameras, but were reluctant to let us use them when we visited technical installations. Since the principal purpose of going to the U.S.S.R. was to learn about its 500-kilovolt electric transmission technology, the trip would be largely wasted if we were prevented from learning important specifications. Our own experts were not ignorant of 500-kilovolt technology; only the details mattered. At the time of the meeting, Udall was on a mission of his own to meet with Secretary Khrushchev, who was then vacationing in the Crimea. Therefore, I was senior in rank, or so we all thought at the time. When the occasion arose for the Americans to speak up, I responded for the group.[3]

I pointed out that learning from the U.S.S.R.'s experience with UHV (ultra-high voltage) technology was one of the main purposes of our trip, that this was a technical mission of energy experts, that several members of our group had been selected solely because of their transmission expertise, that our interest had been fully explained when our trip was arranged, and that there had been no indication from the Soviets that a limit would be imposed on picture taking. The minister said that there was no express provision for unlimited photography in the agreement. I responded by saying that our visit was intended as the start of a planned series of technical exchanges; the U.S. would be compelled to apply the same rules on the Soviets when they visited America. I had no authority to make such a statement, and I doubted that the U.S. would insist on strict reciprocity, but the point had merit and the minister did not know I was bluffing. He suggested a short recess. When he came back, he said we could take pictures freely. Furthermore, if we had any questions at the end of our trip, we could submit them to him in writing.

What could account for this turnabout? It took no intellectual giant to understand that the Soviets had much more to learn about industrial technology from us than the other way around. We did not think, though, that the minister had reversed the Soviet position on his own authority. It might be that in the Kremlin hierarchy there were others who had advised him. My guess was that he had called Khrushchev. Granting permission to take pictures of transmission hardware could only come from the very top.

Once the Russian guides and experts knew that they were safe to do so, they opened up fully and cheerfully. There were no further disagreements, except the small contretemps on the helicopter pad at the "Eisenhower Villa" near Lake Baikal, a matter I shall come to in a moment. Our experts took up the offer of the minister of energy to provide answers to questions submitted in writing. A long list was prepared and sent to the minister. In a short time we received complete and helpful answers. Because I believed the whole industry and its customers were entitled to the information, I sent a copy of the questions and answers to *Electrical World*, a widely read publication in the electric utility industry. It published them in full.[4]

On the trip, our meals were never hurried. There was usually plenty of caviar and always an abundance of vodka. The Soviets seemed to enjoy these occasions thoroughly, perhaps because a surfeit of good food and vodka was not standard fare for them. There were numerous toasts, usually with a chauvinistic tilt on their part, but with the required deference to the themes of international friendship and common national interests. Every toast was celebrated by the Soviets with a shot glass of vodka. The Americans drank more sparingly because of a lack of practice in drinking contests and because they felt unable to compete with the Soviets in this sport. In responding to their toasts, I avoided any boast of American wealth or productivity. Instead, I focused on the features of American life they could appreciate, such as the 60 percent or more of American families who owned their own homes and the almost universal availability of electric service. They did not seem surprised by these data. Perhaps they assumed that our statistics were as fake as their own.

The lavishness of our meals had a bit of the Potemkin Village character. One night Lee White and I walked a little after dinner, then played billiards for a while in the villa billiard room. I believe this was at Irkutsk, and the fantasy quality of two friends of TVA days playing billiards in a villa deep in Siberia was not lost on us. When we finished, our appetites restored by the exercise, we thought we would raid the refrigerator in the kitchen. It was stark empty. No staples, no leftovers, no anything. Apparently, every item needed for a lavish meal was brought in for the occasion. Everything remaining was either removed or eaten by our hosts.[5]

Another example of the Potemkin Village touch was at the great Bratsk Dam. The main hall was a model of cleanliness and polish. The long row of turbine heads appeared from a viewing platform to extend for a mile. When we took a back stairway to the turbine floor, however, all was changed. Some of the steps were treacherous and all were rough in finish. We marveled that there could coexist within the same structure two such contrasting standards of construction.[6]

Our visit to Lake Baikal stands out in memory. We were housed in a villa that was built for an expected visit by President Eisenhower. For his convenience a helicopter pad was located near the villa. On the eve of the visit, the Soviets shot down a U.S. reconnaissance plane in U.S.S.R. airspace. Eisenhower refused to take responsibility for the "spy" plane or to apologize. Khrushchev promptly canceled Eisenhower's visit. The villa was a little larger and more comfortable than the others, and we enjoyed the thought of living in a facility denied to Eisenhower. About a mile away, up a steep hill, was a point that offered a dramatic view of the lake and the surrounding hills. In climbing the hill, Udall led the way, at a clip midway between a fast walk and a hard run. The other members of the party straggled behind and, at a considerable distance, I brought up the rear. Stewart was then in his early forties, strongly built and in excellent physical condition.[7]

The same or the next day, our hosts declared an intermission during an exceptionally hearty lunch; they wanted us to complete the meal with the same zest and dedication with which we started. Udall decided to use the interval for more climbing. The rest of the party, myself excepted, elected to go with him. While I was trying to decide whether to tag along, our constant companions, Poliakov and the young woman I have mentioned, asked me if I would like to join them in a short walk. I accepted gladly, and the three of us strolled for a few minutes in the sunshine of a beautiful day, winding up on the helicopter pad, which was otherwise deserted. Then began one of the stranger incidents of the trip. The two KGB people soon changed the conversation from small talk to a description of the glories of the Soviet Union and communism. The atmosphere became tense. They talked about the egalitarian foundation of the Soviet Union, its great economic achievements, and the pride of its people. I could not let all this pass, so the conversation became a debate. I mentioned Stalin's son, a general in the Soviet Army, and asked whether this came under the heading of egalitarianism. I brought up several other incidents, now forgotten, of well-known favoritism to high bureaucrats. I told them that I would not quarrel with their claims of economic progress, but that in the United States so many people were sated with material things that they were seeking higher values. (Remember, this was in the Kennedy years, when it was no disgrace to mitigate greed with a trifle of idealism.) I told them that I was the son of Russian-Jewish immigrants. I asked them to consider what my life prospects might have been had I lived in the Soviet Union. I also told them something about the broad distribution of income in the United States. By now the KGB team had given up their proselytizing, but I was still going strong. The bald effort to recruit seemed insulting, and I showed my resentment.

At this point, the exercise party returned and came to the pad. Udall asked what all the shouting was about. I told him we were having a little debate on the relative merits of the U.S. and Soviet systems. I had a camera with me and took the occasion to snap a picture of our two KGB friends talking to Udall. When I returned to the U.S., I gave the snapshot to the FBI, as well as a copy of a diary I had kept on the trip. I had not made the trip for the FBI's benefit, but KGB efforts at recruitment were not covered by any privacy consideration.[8]

Reflecting on the helicopter pad incident after many years, I must say that it is possible I contributed to the KGB's attempt to recruit me. I brought with me on the trip some small gifts, such as a roll of dimes and quarters, cigarettes, a copy in English of a Dostoyevsky novel, and a white silk U.S. aviator's scarf I had bought at one of the Smithsonian craft shops. The book, a paperback, should not have raised eyebrows, even in the U.S.S.R., but the local official to whom I gave it probably shared the American spy pathology, may never have read a book except for Lenin's writings, and perhaps had not heard of Dostoyevsky. I gave the scarf to another official. It was not an ordinary white silk scarf, but a special World War II Air Force issue. It had a map of the Soviet Union printed on it so that a downed aviator could find his way to a place of refuge. Certainly it had no direct combat uses—no targets, no military information—just enough data for orientation in a strange environment. To my surprise, the official gave no expression of gratitude. He accepted it in silence, almost reluctantly. I thought he did not understand, but it may have been more than that. Perhaps he thought I was trying to compromise him. On the other hand, he or the KGB may have thought the gift was a coded signal to the KGB. I suppose now that he delivered the scarf to the KGB as soon as possible. There is no way to know what he or the KGB made of the scarf, but my guess is that they found some sinister possibilities in it.[9]

At one of our visits we were treated to a cartoon movie comparing U.S. and Soviet rates of economic growth. Among other comparisons it showed two little trains, theirs and ours, on parallel tracks, in a race to expand electric power capacity. The Soviet train was behind the American train, but was rapidly catching up. The narrator predicted that the Soviet Union would surpass the United States by 1980. Other comparisons suggested similar conclusions. Since we knew something about the Soviet penchant for creative statistics and overly sanguine forecasts, we were not impressed. My line, as usual on the trip, was that we wished them well in their economic program, that we were not in competition with them, that we were satisfied with our own rate of economic improvement, and that we were increasingly concerned with a better distribution of income. This latter point could be made seriously in the Kennedy-Johnson years.

The London Trip

Another trip I made was to London to speak to a large group of European investors on the attractions of electric utility equities. The trip was conceived and paid for by Model, Roland, and Company, a New York investment house with Rothschild affiliations. The bankers believed that the merits of American electric and gas stocks were not adequately appreciated in Europe. The United States was then the only large industrial country with a primary reliance on investor-owned electric and gas companies; the British began a process of privatization of their gas and utility industries only in the late 1980s. If investors wanted a stake in the electricity or gas business, the U.S., as a practical matter, was the only source. Investors abroad were chary of American utility equities because of their fear that the federal or state governments would expropriate the [utility] companies. If they fully understood the safety of American utility securities, the U.S. bankers argued, they would buy them in larger volume. This would add to the pool of capital on which U.S. companies could draw, reduce the cost of capital, and ultimately benefit ratepayers. Anything for the ratepayers, I thought, even at the sacrifice of making my first trip to Europe. In self-defense, there was a solid business purpose, and I did not linger.[10]

Carpenters Hall [where Swidler gave his talk] was well filled with perhaps 150 investment people. I must admit my talk was uncommonly humorless and dull. Perhaps I was assuming that people who spent their lives reading financial prospectuses were beyond appreciation of the light touch. At any rate, I explained the unique regulatory structure of the United States, the division of functions between the federal government and the states, the constitutional roles of the executive and the judiciary, the structural differences between the gas and electric industries, and much, much more. Coming and going, I explained how judicial oversight of the regulatory process assured substantive due process in protecting investor interests. I also described the solid foundation and glowing prospects for both industries. Some of my listeners may have closed their steely eyes for a nap while I spoke, which I would have done in their place. Did I open up new markets for utility securities? Did the bankers recoup the costs of the trip? I shall never know. They were complimentary at the time.

I could not make a talk of the same tenor again. In substance, what I said was correct at the time, but the times have changed. The early 1960s was a very stable period in the history of the two regulated energy industries. Prices were declining slightly in current dollars, and more significantly in constant dollars. The electric industry was doubling in size every decade. State regulation was lax, and even the FPC understood the need to allow earnings high enough to attract the enormous amount of new capital required every year. The electric industry was dipping its toe into nuclear technology. The manufacturers of nuclear plants were offering turnkey contracts. Little, if any, consideration was given to environmental problems. In another decade, remarkable changes took place, enough to shake confidence in the ability or willingness of some of the state regulators to allow earnings adequate to enable the utilities to build for the future, or even remain solvent.[11]

CHAPTER 24

Leaving the FPC, 1965

In the spring of 1965, with my term expiring in June, I began to think seriously about my options for the future. Did I want a second term at the FPC? Or should I re-enter private practice. If so, in Nashville or Washington? If Washington, should I join one of the large firms, or open an office of my own? I had good partnership offers from my former Nashville associates and from a large Washington firm, both unsolicited. After some threshing of the alternatives with Gretchen, I decided that I would stay on if Johnson wanted me. Otherwise, I would open an office of my own in Washington. Gretchen said she thought I would be happier continuing with my work at the FPC; the prospect of greater earnings in private practice should not be a factor. Ann and Mark chimed in on the same note.

Johnson had a summary way of disposing of Kennedy appointees. He wanted his own people in key jobs. He remained close to the oil producers in Texas, who were strongly opposed to my reappointment. I knew reappointment could not be taken for granted. Having witnessed two of my colleagues lingering for months after their terms expired, with Johnson considering their fate, I had no desire to share the holdover experience. I decided that I would discover the president's intentions in order to make other plans if he preferred someone else for my job.

John Macy was then serving in a dual role for Johnson. He was chairman of the Civil Service Commission and the president's chief headhunter and personnel adviser. In March of 1965, at a luncheon meeting, I asked his help in clarifying my situation. I told him my term was coming to a close, that I should be glad to accept reappointment if the president wished, but that I could not wait until the last minute to find out the president's intentions. If I was not to continue at the FPC, I intended to enter private law practice. I would need to rent an office and make other arrangements. I added that if the president indicated he did not wish me to stay, I would be unavailable if he should change his mind. Macy agreed to sound out the president. He called me back in a day or two. The president understood my position and said that I should feel free to consider other opportunities. The message was short and clear. I proceeded with my own plans.[1]

It soon became apparent that one change in detail would be necessary in these arrangements. The Permian Basin case, the landmark proceeding on fixing natural gas rates on an area basis, could not be put to rest by the commission (a petition for rehearing was pending) for some months after my term would expire in June. I did not want to stumble on this crucial ratemaking innovation, and I presume Johnson did not want to risk embarrassment if the commission changed course in my absence. When I talked again with Macy it was agreed that I would stay on until November 30. David Freeman, my executive assistant, agreed to enter a partnership with me, and we soon signed a lease (effective December 1, 1965) for offices at 1750 Pennsylvania Avenue.

I made no public announcement of my plans, but word soon leaked out. The president began getting letters from a variety of sources urging my reappointment. The public power people regarded me as one of them, and they made their support known. The utilities, strangely, also urged reappointment, not because I was their first choice, but because it was unlikely that my successor would be an improvement from their perspective. Before these voices were heard, I assume what President Johnson knew about me came mainly from the producers. They had good reason to oppose me. The field prices for natural gas determined by the FPC were well below what they considered reasonable. I had a small personal acquaintance with the president because Lynda Bird, his elder daughter, and my daughter Ann were classmates and good friends. Gretchen and I and a few other parents had attended several small parties at the Johnson home, but these social settings provided little basis for appraisals of my job performance.

At any rate, sometime in the late summer of 1965, Lee White told me that the president would reappoint me if I were willing. As I noted earlier, Freeman and I had already rented offices and made other arrangements for setting up a law firm. Besides, I had told Macy explicitly that I would not be available if the president changed his mind. I explained to White, with due expressions of regret, that I could not accept. I reported this conversation to Freeman, and he did not disagree. He believed he had nothing further to gain by continuing at the FPC.

The offer of a reappointment was not the end of the story. When the Northeast blackout occurred in late November, the president asked me to stay until the end of the year. With Freeman's concurrence, I readily agreed. The president was able to postpone my lease obligation with no trouble—the magic of a call from the White House. Apparently, my handling of the Northeast power failure removed any lingering uncertainties the president had about me. During the course of the investigation, I was in steady communication with the White House. The public reception of the investigation and subsequent report reflected favorably on the president. The White House attitude perceptibly changed for the better. When I told Bill Moyers that the report was ready for submission to the president, he asked me to hold it for a few days until an appropriate occasion could be arranged. The president was then at his Texas ranch. His chief financial advisers were scheduled to visit him there. I was asked to join them on the trip, present the report to the president, have lunch with the president and Mrs. Johnson, and proceed by helicopter to Dallas for a press conference. I asked to take Stew Brown with me, and he was included in all the arrangements. When we arrived, the president took all the visitors on a trip around the ranch in his open car. His enjoyment of the ranch was plain; he talked the whole time, calling attention with pride to the fine condition of his cattle, the quality of the fencing, and the excellent grass cover.[2]

The first agenda item was the presentation of my report, which took place outdoors, and was covered by photographers and television cameras. The president deferred to me for a short talk on the blackout investigation and the report. He confined himself to a few gracious remarks about my work. At lunch, which started with a brief prayer by the president, the food was simple but good. Shortly after we finished the meal, if I recall the sequence correctly, the president received a report that changed my schedule for the rest of the day. Of all things, a large area in western Texas and some adjoining states had suffered a blackout. Johnson's own communications at the ranch were affected. The president asked Stew and me to make an on-the-spot investigation. A well-equipped communications center was attached to the house. The Dallas press conference was canceled, and Stew and I went to work. Stew had been asked to join me on the trip as a reward for his important contributions to the investigation and report, but it proved lucky indeed that he was available for a tense afternoon. It was he, principally, who was able to define the area of trouble and the causes (this time there was no mystery). Before we left in the late afternoon to return to Washington, Stew was able to give the president a clear picture of what had happened and the status of recovery (which was rapid).

While I was at the ranch, Johnson drew me aside to say that he wanted me to stay at the FPC or in some other post. I told him I could not do so. I waited expectantly for some of the celebrated Johnson arm-twisting, prepared to turn down anything short of a Supreme Court appointment, but somewhat to my disappointment, it did not come. He next tested me out with an offer to serve as the first chairman of the new administrative conference in the Department of Justice. I had no difficulty turning it down. Quite aside from any innate attractions it may have had, it was a position better suited for a professional scholar than for me.

In a book Dave Freeman [later] wrote for the Ford Foundation, he bracketed me with two other commissioners, stating only that I had not been reappointed to the FPC. The general point he was making was that consumer-oriented commissioners are rarely reappointed. Whatever the merits of this argument, he was unfair to Johnson in his reference to my case, since Johnson had offered me reappointment, if only on second thought. Freeman was also unfair to me by implying presidential dissatisfaction. Johnson had shown me many evidences of friendship and high regard.[3]

Some months after Freeman and I had opened our law office, I was given an offer that required serious consideration. The chairman of the AEC, Glenn Seaborg, a celebrated physicist, came to see me. He said that there was a vacancy on the AEC. With the knowledge and approval of the president, he asked if I would accept an appointment. Seaborg and fellow commissioner Jim Ramey, an old friend of TVA days, were seeking someone who would take responsibility for the reactor licensing program, which was not going well. He and Ramey believed that I was the right man for the job. To my later regret, I turned it down.[4]

The job had great attraction. Yet there were several fair to middling excuses for rejecting it. I was beginning to make money again after a long drought. My clients would be inconvenienced. The offer was for a commissionership, so I would lack the freedom and the prestige of a chairmanship. I knew little of nuclear technology, just enough to appreciate its dangers. These were the lesser reasons for rejecting the offer. There were, of course, counter-arguments as well. The salary (higher than those of FPC commissioners) was adequate for a comfortable life. My clients, mostly electric companies, might be glad to have me try to straighten out the licensing mess. The prestige point would not even have occurred to a better man, and I was a little ashamed that it surfaced in my mind. After all, I was not at the beginning of my career. At the age of sixty I was not looking for stepping-stones to higher office. The AEC had a practice of delegating responsibilities, which meant that I might have had an adequately free hand as the commissioner in charge of licensing. The lack of in-depth knowledge about reactor technology was not fatal. I would have had time to study and ask questions. The offer was not based on my ability to build or operate a nuclear reactor, but on my demonstrated capability in the licensing field.

The ineluctable reason that compelled me to walk away from the offer was that to do so would have left Freeman in the lurch, or so I thought. He had not yet developed his own clientele. I had brought in all the initial business, much of my work was personal, and it was uncertain how many clients would stay with the firm if I left. Without careful weighing of the other factors, I turned down the AEC appointment. Of course, I told Freeman promptly of the offer and my rejection, but did not say that protecting him was a factor. If I had, he would probably have felt compelled to resign from the firm, and that would have been unfair. Ironically, a little later Freeman resigned to take a post as an energy consultant in the Bureau of the Budget during the Nixon Administration. By that time the AEC vacancy had been filled.

Rejection of the AEC offer has stayed in my mind as a constant source of regret, though my sense of having missed an opportunity set the stage for my interest in the next offer of a government job.

PART VII

Back to Private Practice, 1966–1970

Except for a few bumps, Swidler's return to private legal practice went smoothly. He attempted to remain in contact with his FPC colleagues, but quickly realized that maintaining those relationships risked too many conflicts of interest. Turning his attention elsewhere, Swidler focused on enlarging his law firm. Not surprisingly, some of the people he hired were acquaintances from the TVA and the FPC.

In addition to serving former clients, Swidler also worked on acquiring new ones. None of the latter presented more challenges than Consolidated Edison of New York ("Con Ed"). A major supplier of electric power to the New York City metropolitan area, Con Ed, like the FPC before Swidler arrived, had a reputation for poor management. Con Ed followed Swidler's advice and hired Charles Luce as the company's chief executive officer. In his new position, Luce often sought Swidler's counsel on key issues, especially in regard to restoring consumer confidence. Swidler's own experiences and integrity shaped the suggestions and recommendations he offered.

Ironically, it was during this period of private life that Swidler encountered cynicism among public officials. He describes one particular episode involving the administration of Mayor John Lindsay. Though Swidler's memory had begun to fade when he recalled the incident, enough details remained to leave a sour taste in his mouth. Here, as elsewhere in his memoir, Swidler expresses disdain for those who would exercise power solely for political advantage. He realized that every public official has to make certain calculations, but he also believed that the aims of public life are distorted when those calculations supersede all other considerations.

CHAPTER 25

A Law Practice Interlude in Washington, D.C., 1966–1970

On January 2, 1966, Dave Freeman and I opened our shop for the practice of law. I had turned down an offer from one of the Washington megafirms that wanted me to introduce an energy practice into its already diversified mix of business because I preferred the freedom of a small partnership.

Despite my awareness of the need for regulatory officials to maintain their distance from practitioners, I still felt part of the FPC. I sought to maintain the old relationships with my former colleagues; they had been good friends and I did not want to lose them. For four-and-one-half years, much of my life had been spent in their company; my relations with them anchored a lot of my thinking. I am glad to say that they promptly set me straight. Only at the first meeting, at a luncheon in my office, was there the old informality. My next invitation to them was received coolly. While I was hurt at first, common sense finally took hold and I realized that these relationships could not survive the potential conflicts of interest. I did not try again. If someone else had been in my position and had asked me for advice, I would have told him, without need for reflection, that close personal friendships between commissioners and practitioners are unhealthy. In my own case, it took a little time to realize that.

The first client of the firm was Commonwealth Edison Company of Chicago, which I had fought thirty years before when it was a part of Samuel Insull's empire. Its chairman and chief executive officer was Harris Ward. He had been a member of the EAC of the National Power Survey.

Other clients followed—Northeast Utilities, Dow Chemical Company, Pennsylvania Power & Light Company, Consolidated Edison of New York, Pacific Lighting Company of California, and the Acres Company, to name a few. I declined to serve one large power company because I suspected the soundness of its business ethics. In a large law firm, turning away a large company against whom no wrong had been proven, or even alleged, would have been unthinkable. I had no solid reason for my opinion, and even less basis for sitting in judgment. But I enjoyed the freedom of choosing my clients, even though I knew that turning away potentially valuable ones was a poor way to build a rewarding practice. I believe now that I was wrong in jumping to conclusions. I may have deprived myself (and the firm) of interesting work and the pleasure of interactions with compatible people.

Dave eventually resigned from the firm. Before he left, he had suggested that we bring in Richard Freeman (no relation), who had served on my staff at TVA. I was quick to agree because I knew Dick to be a good lawyer and a fine person. I also liked him very much. He was a partner in a small Chicago firm headed by Noel Belnap, a formidable figure in transportation law. Dick accepted our offer and made an additional proposal: He advised us to merge the two firms. Since the Chicago firm specialized in federal transportation law, and our firm specialized in the energy area, and because of the problems created by physical separation, I did not see much synergy in the proposed arrangement. The enthusiasm, though, of the two Freemans, who had become close friends in Knoxville, won me over.[1]

The merger had little value for the Washington end at first. It created much paperwork and tedious long-distance telephone conferences. Two of the Chicago lawyers soon moved to Washington and became partners. Belnap retired and died soon after.

When Dave Freeman resigned, I invited Charles J. McCarthy, a close friend and TVA colleague for many years, to join the firm as a partner. McCarthy accepted the invitation. He was an exceptionally able lawyer and a talented litigator of much experience. While working for the TVA, he had participated in all the important energy and constitutional law cases. He had also supervised the growing workload of TVA nuclear licensing. The firm soon recruited William Harkaway from the FPC. He was a litigator and an expert on natural gas problems, one of the brightest of the FPC staff lawyers. With these two on board, we were better equipped than ever to handle the firm's mix of business.[2]

Private Clients: Chuck Luce and Consolidated Edison

My most rewarding work, I believe, was for Consolidated Edison of New York. Often referred to in New York City as "the company you love to hate," it was the butt of ridicule. In the electric utility industry it was considered a black sheep and a threat to the reputation of well-managed companies. The Board of Con Ed recognized, finally, that for too long it had tolerated poor management. A revolt led by Frederick Eaton, a Wall Street lawyer of high reputation, resulted in the decision to seek a new leader to salvage a company on a downhill slide.[3]

John Corson interviewed me [in early 1967] and asked me for suggestions. I mentioned the names of a couple of promising executives in other companies. After some inconsequential discussion of these people, in whom Corson showed little interest, he uncovered one of the cards in his hand. How about people in public life who might have the qualifications to lead Con Ed? I could think of only one man in this category who was not barred by approaching retirement age: Chuck Luce. I knew Luce well. Our friendship predated his appointment as chairman of the Bonneville Power Administration. At the time of the Corson interview, he was the undersecretary of the interior, the second ranking post in the department. Corson questioned me closely on his character, career, attainments, and leadership abilities.

Luce asked me to help him after he accepted the position. During his first few months on the job, our collaboration was close. I spent many days in Con Ed's offices as a sort of general consultant. At one of the earliest meetings, perhaps the first, the question of publicity came up. I strongly supported an effort to put a new face on Con Ed by letting the public (including Con Ed's employees) know that in the future Con Ed would be a very different company. The medium should be full-page ads in the city's major newspapers, one a week in an initial period of a month or two. I also urged that the ads be in the form of signed statements by Luce. Luce was not anxious to intrude personally; after some discussion, however, he was persuaded that his name and picture would help in convincing the public that the change in helmsman meant a real change of course. The first of the ads was outlined and partially written at that meeting. It was matter-of-fact in acknowledging the company's faults and in setting out future goals and procedures. The series was well received and helped launch the Luce revival of Con Ed.

When I came to the FPC, the electric rate work was being handled as an orphan group within the gas rate staff. In Con Ed I found the mirror image of the FPC situation. Con Ed's gas business was only a fraction of its electric business. It was managed (poorly) by the electric staff, who were not overly capable on electric matters; they were downright inefficient on the gas side. Many pipeline and producer cases in which Con Ed was not a necessary party established precedents that would affect distributors generally, and Con Ed in particular. Considering the magnitude of its stake, Con Ed was one of the least active and effective advocates for its causes. I called this situation to Luce's attention and he soon put an end to it by setting up a separate, and sharp, gas staff.

Con Ed, at the time of Luce's appointment, was desperate for additional power sources. Luce asked me to help in solving this problem. Possible sources included two of Con Ed's closest neighbors, Orange & Rockland Utilities and Central Hudson Power & Light, both of which were then building large oil-fired stations. Negotiations had already started with these two utilities, but they had lapsed, either because of failure of prosecution or because of disagreements that the negotiators could not bridge. Assisted by the old Con Ed negotiating team, I took over these discussions. I brought them to a reasonably early conclusion on a basis more favorable to Con Ed than had previously been discussed—larger blocks of power for Con Ed, reduced and delayed withdrawal rights by the sellers, among other features. The two agreements went far in resolving the immediate power supply problem.

In his book, Luce gives me credit for working out a compromise between Mayor John Lindsay and Con Ed after I became chairman of the New York State Public Service Commission. The compromise permitted construction of one additional Con Ed unit within the city (at the Astoria site), but on the assurance that it would be the last one built in the city, and that it would not contribute to air pollution. I have no recollection of this episode; Luce, however, kept notes and I did not. I am glad to accept credit.[4]

While I do not remember the negotiation with Mayor Lindsay about the Astoria unit, I recall another discussion with him and his people about the so-called "aqueduct" transmission line. As it happened, Luce, in his first six months, had uncovered a criminal bid-rigging conspiracy among paving contractors, including a company owned by Henry Fried. Fried's company was taken off Con Ed's approved contractor list and denied all future contracts. Subsequent investigations revealed that a Con Ed vice-president, Gerald Hadden, was on Fried's payroll at $1,000 a month, plus $5,000 as a Christmas present. These investigations also revealed that Hadden arranged to pay a bribe of $20,000 through Fried to James L. Marcus, the New York City commissioner of water, gas, and electricity. The payment was apparently made to secure, from Marcus, renewal of the permit needed by Con Ed for maintenance and expansion of the aqueduct transmission line. Marcus and others (including the contractor) were convicted. The whole incident was known as the "paving scandal."[5]

I was not directly involved in the Con Ed response to the paving scandal. I mention it here for the light it throws on the ethics of prominent members of the Lindsay administration. Instead of granting a renewal of the permit to Con Ed for the original fee, the city demanded nine times that amount, on the basis that the rebuilt line that Con Ed was proposing would carry nine times the amount of electricity. The improvement was important for reliability of service in New York, yet Con Ed was compelled to pay a half million dollars more per year for a permit that could have been obtained for a bribe of $20,000. Clearly, honesty did not pay in dealing with the city. In the end, of course, it was an important part of the successful effort to transform the Con Ed reputation, to earn respect, and to impose standards of economy and efficiency upon that organization.

My participation in this matter consisted of a meeting, while I was still in private practice, with Mayor Lindsay and some of his key aides in an effort to persuade the city administration that working with Con Ed, instead of using it as a political football, would yield large benefits to the city itself and to its citizens. I also took the opportunity to explain the urgent need for a permit to rebuild the aqueduct line. Mayor Lindsay, as I recall, did not respond directly. He smiled, but not warmly. After a bit, he excused himself and left me with his aides. One of them, a senior political adviser, acted as spokesperson. I pressed him a little with the argument that Con Ed was carrying out, by delegation, an essential municipal function. Furthermore, cooperation would enable Con Ed to reduce costs and improve service. There was silence for a few minutes. Finally, the spokesperson responded, frankly and in the plainest terms, that Lindsay could not afford to make peace with Con Ed; the company was so unpopular that it was politically profitable for Lindsay to be able to "kick Con Ed around."

Lindsay was handsome, telegenic, charismatic, and had lofty political ambitions. He miscalculated, though. A little cynicism may be necessary in politics, but not when it distorts reality. Lindsay carried it too far, too soon, and too ineptly for his own political future. He was not re-elected. Other politicians in New York and elsewhere have been much worse, without suffering so badly, or at all. Lindsay himself was not implicated in the bribery that took place in his administration. He lacked, however, the political skills to maintain an image of seriousness and probity in carrying out the public's business. The appearance of piety is an almost essential part of a successful politician's baggage.

For reasons of client privacy, I can hardly divulge much of significance about other matters. The life was pleasant and financially rewarding, but a bit boring. It did not possess the challenge and excitement of my FPC work. It lacked a goal more important than advancing the sometimes doubtful causes of private clients. I had been spoiled by long government service. Supporting my family came first, of course, and if I had had no other options, I would have been content in private practice. I still yearned, though, to make a difference in public life.

PART VIII

Chairman of the New York State Public Service Commission, 1970–1974

Despite wanting to "make a difference in public life," Swidler seemed unimpressed with Governor Nelson Rockefeller's 1969 offer to become chairman of the New York State Public Service Commission (PSC). Accepting the position would mean relocating to another state and facing the inconvenience of a reduced income. Furthermore, the PSC was an agency whose reputation was abysmal. Democrats and other groups had condemned it as a sinecure for Republican politicians, while the New York State Commission of Investigation had criticized it for not holding down utility costs. For these reasons, Swidler might have declined Rockefeller's offer. Nevertheless, given his enthusiasm and proven ability to meet various challenges, he might have viewed these circumstances as enticements, not just disincentives. Whatever the case, he took the position. He and Gretchen moved to Albany, New York, where they spent the next four years.[1]

Consistent with the reforms he implemented at the FPC, Swidler focused first on making structural and organizational changes within the PSC. As he noted in his postscript to the PSC's annual report for 1971: "This is the story of a Commission in flux . . . of a Commission undergoing drastic changes to maintain the ability to deal adequately with the problems of power, communications, and water supply." Up until Swidler's arrival, the PSC's staff had been divided, somewhat illogically, between Albany and New York City. Swidler put an end to this arrangement by moving the PSC's New York City personnel to Albany. Moreover, he gave the staff an augmented role in PSC proceedings. Both of these changes increased the commission's effectiveness. Additionally, Swidler created an economics and an environmental division within the PSC. The former improved the commission's ability to assess complicated rate structures, while the latter enhanced its environmental oversight functions. Swidler also made key personnel changes, including the appointment of the first African American woman to the commission.[2]

The most serious problem Swidler and other regulators faced was not organizational, but economic. The early 1970s witnessed the phenomenon of stagflation, a combination of stagnant growth and high inflation. One cause of the era's inflation was a dramatic increase in energy costs, which was a consequence of the rise in power and influence of the Organization of Petroleum Exporting Countries (OPEC). Between 1973 and 1974, OPEC was responsible for a fourfold increase in the price of petroleum. Even prior to that increase, American utility companies were feeling the effects of rising costs. During 1970, thirty-six rate cases requiring hearings came before the PSC; in the twenty-one cases that were decided before the end of the year, the PSC awarded increases totaling $264 million (out of a requested $370 million). Determining the precise amount to grant for each request, however, was not a simple process. Regulators had to figure out how much of a requested increase was actually linked to inflation, and how much might be an attempt to accrue unfair profits. Even when an increase was justifiable, regulators often faced public criticism for permitting higher rates. More problematic still, higher utility bills can themselves become part of an inflationary spiral. Swidler was aware of these dynamics. He addressed them by relying on the lessons he had learned at the FPC, and by implementing a pragmatic, innovative approach to problem solving. Though not successful in every matter, he had some notable victories.[3]

On another issue, environmental protection, Swidler's reflections are qualified. As noted above, he oversaw the creation of an environmental division within the PSC. He also helped craft environmental legislation that gave the PSC control over the placement of electric and gas transmission lines. Yet Swidler also thought that his own staff sided too quickly with conservationists in their battle against the utilities. He did not believe that the PSC's staff recognized the proper "mix" of environmental and economic considerations necessary to advance the public good.

The challenges and frustrations Swidler faced at the PSC prompted his prescient call for a national agency to coordinate the country's energy resources. In 1971, during the course of congressional testimony, he remarked: "It seems abundantly clear that no state or group of states, can solve its energy dilemma. I believe that the only solution is to create a federal agency with the authority to formulate and apply the policies necessary for the supply, allocation, and use of all our energy resources, including policies which would direct our research programs toward projects that offer the best hope of solving the energy problem. Such an agency must not only have authority with regard to energy demand and supply, but must also have an important voice in setting policies for national security, environmental protection, land use, and economic growth."[4]

This statement not only revealed Swidler's impatience with the limitations of piecemeal regulation, it also indicated his continuing faith in the efficacy of the federal government, a faith likely fueled by his experiences at both the TVA and the Federal Power Commission. In the fall of 1977, with the establishment of the U.S. Department of Energy, Swidler's plea for a national energy agency became a reality.

CHAPTER 26

Working for Rockefeller in Albany, 1970–1974

One day in the fall of 1969, a few days before Gretchen and I were scheduled to leave on a New England vacation, a pleasant young man named Harry Albright came to see me at my office in Washington. He said that, at the direction of Governor Nelson Rockefeller, he was seeking suggestions for a person with the qualifications to serve as chairman of the New York State Public Service Commission (PSC). I came up with a couple of names. We discussed briefly their fitness for the job and possible availability. Then Albright disclosed that the real point of his visit was to see if I was interested in the job. I told him that I was not available, that he must be under a misconception to make the offer. I said that since leaving the FPC I had developed a clientele consisting mostly of investor-owned electric utilities, including Consolidated Edison of New York City. Also, I was a Democrat.[1]

Albright explained the New York situation more fully. Governor Rockefeller would be running for re-election in November of the following year. The PSC was a weak spot in his record. The incumbent chairman, a political figure, had lost the trust of the legislature, consumer organizations, and the press. Rockefeller had instructed Albright to find the best man for the job, regardless of party; he preferred someone well and favorably known whose appointment would assure voters of competence. Albright was an expert at massaging egos, and, of course, I was pleased by the compliments. Nevertheless, I did not see how I could give serious consideration to the offer. He asked me to think it over, and I said I would. In a telephone conversation a day or two later, I told him that I would call him when Gretchen and I got back from our New England vacation.

Albright called me again, perhaps the next day, and asked whether on our way to New England we could meet in New York City; I agreed. He met us at our hotel. He said Rockefeller was anxious for me to accept and would try to accommodate any reasonable request. He provided a better picture of pay and emoluments, which included a car and driver. As for the conflict of interest problem I had raised, the governor was satisfied that I would be judged on the basis of my whole career, not merely on my current work. Rockefeller had talked with an editor of the *New York Times*, who was reassuring. Albright asked for my commitment. I declined, but we did talk about some of the practical problems if I should take the job. Before declining, he said, I should at least meet with the governor. I said I would get in touch with him after our vacation.

The detailed bargaining began after we returned to Washington. I met with the governor personally at his New York office. I told him what I needed from him in order to do a good job. My first request was that I be able to talk with him whenever necessary. He said he would always be available to me, and would so instruct his secretary. I outlined the kind of legislation that would be essential. It would have to: (1) create a system that would permit the chairman to be a strong leader; (2) organize commission appointments on a bipartisan basis, with no more than three of the five commissioners from the same party (this would allow me to serve as a Democrat); (3) place the responsibility for the licensing of generating plants and transmission lines with the PSC, including a pivotal role on environmental questions; (4) give the [Environmental] Conservation Department an important advisory function; and (5) include a reasonable basis for rate awards to regulated utilities. I also asked for a sympathetic ear on manpower and related budget questions. To all this he agreed.

In turn, Rockefeller (who asked at this first meeting that I call him by his first name) said that he was on record to transfer the PSC's jurisdiction over rail and bus transportation to the Transportation Department. The transfer was scheduled to take place a little more than a year after I was to take office. Did I object? I said that since I thought transportation involved a different set of problems than utility jurisdiction, I would readily agree to the transfer. He told me that the PSC had status as a cabinet agency. He had under consideration a proposal to amend the state constitution to use the cabinet slot for another agency. I said rather stiffly that I should not like to see a loss of status for the PSC during my tenure. He said he would not recommend the change. He also warned me that while he would recommend to the legislature the proposals we had discussed, he had many other legislative irons in the fire. It would be up to me to take responsibility for getting my bills passed. He said I should cultivate the friendship of the legislative leaders in both parties. On the basis of these terms, we shook hands.

A few days later the governor sent me a draft of a proposed public announcement of my appointment. It suited me well enough, except it failed to mention that I had represented Con Edison. I said that the announcement must include a disclosure. The governor's representative said that Rockefeller did not think this was an auspicious time. I said it was the only time. If there was to be strong opposition, the announcement would flush it out before a serious mistake was made; if the information leaked out afterward—and it could not be kept in confidence forever—it would appear that the governor and I had withheld crucial background material. The governor finally yielded and the disclosure was made. It elicited neither criticism nor comment. After I had been in office a couple of years, a reporter phoned me in great excitement to verify a rumor that I had been a lawyer for Con Ed. I said that not only was it true, but it was disclosed in the initial announcement of my appointment. I added that on taking the job I had broken all ties with my former firm. When he checked and found out that his information was not the basis for a sensational scoop, he dropped the matter.

Rockefeller had many strong points. He was quick of mind, careful and decisive as an administrator, and cognizant of his own strengths and limitations. He was a political professional. He was never pompous, but he knew the special advantages created by his wealth. He expected and received a higher degree of deference than was warranted by his political position alone. He once told me that his mother had emphasized that he should associate with people smarter than himself, for only in this way could he improve his own mind. While it was a flattering thing to say to me, I assumed he had told the same story to many others. He was particularly (and I believe genuinely) impressed with people who had made their own way to high position.

Getting Started at the PSC

It was late November when I agreed to take the New York job. Rockefeller wanted me to start at once, so he put me on his payroll as a consultant. I spent much of the time with the secretary to the governor and the legal staff drafting legislation. We worked on two bills: One concerned the PSC's organizational structure—the agency's procedures were to be streamlined and it was to be given a new role in long-term energy planning—and the other focused on the licensing of new generating stations and transmission lines. I also met with the legislative leaders, introducing them to the governor's program for the PSC.[2]

Since it would have been intrusive to have attended PSC meetings before I took the oath of office, it was not until sometime in January before I could do anything within the PSC. I started by interviewing the division heads and talking with section heads or other staff members at the middle echelon. To my surprise, the electric and gas staffs were combined in a single unit. The head of the unit was a veteran of the PSC. He was something of a tyrant who rejected all staff suggestions. Since he received no instructions from the commission except, perhaps, not to bother them, he did not want, for his own part, to be disturbed by staff. As I phrased it at the time, if anyone on the staff came up with an idea for improvement, he whacked him over the head with a two-by-four (they were all "hims" in that fading period before women asserted their rights). Clearly, this gentleman had to go. His long service entitled him to a retirement income that would spare him from financial hardship. Although he seemed to doubt that the commission could get along without him, he tendered his resignation on request.

I had little trouble in securing agreement on the PSC's organizational changes. Remedial changes that would enable the commission to function more effectively were difficult to oppose, even by those who sensed serious differences with me on substantive issues. I was too old to be suspected of empire building, and it further allayed suspicion when I recommended shortening the commissioners' terms. One change proved especially useful in preventing clogged dockets. It provided for formal hearings in all rate cases, except those involving increases of less than 2 percent a year. In the period of double-digit inflation that ensued, this provision made it possible to handle small increases expeditiously. It also tended to keep rates down. On a number of occasions utilities settled for an early 2 percent increase when, on a fully contested record, the company could have received much more, but much later. All such cases came to the commission after careful staff evaluation and recommendation.[3]

The PSC's employees (some six hundred people in all) were divided geographically. About half were located in Albany. They included the general manager, Sam Madison (ably seconded by Bill Barnes), the personnel people, the lawyers, most of the hearing examiners, the telephone division, and the transportation group. The gas and electric division, the water division, one or two hearing examiners, and about half the accountants were in New York City. This separation meant that a meeting could not be held at either place if it involved people in both locations, except by having individuals go from one place to the other, with the attendant expense and loss of time. As a result, there were few meetings that included members of both groups. Each of the commissioners had offices assigned to him in both places, but they spent little time in either. They met only on the days of formal meetings, which alternated between the two locations. The bus was careening down the mountain without a driver. I took prompt action to clear up these problems.

I sounded out the people in New York and Albany as to their willingness to move. It appeared that most of the city people would move to Albany, a few of them reluctantly. The Albany staff was almost unanimous in rejecting a move to New York City, although some would probably have changed their minds if that had been necessary to save their jobs. The staff preference confirmed my own judgment that Albany was the better location. It would be easier in Albany to keep in touch with the governor and his staff, with the other state departments and agencies, and with the legislature. In a short time the electric and gas staffs moved to Albany.

In addition to filling gaps in existing staff units, I created two new working groups. One was an environmental group. There was no such group at the FPC by the time I left at the end of 1965, although there were several people who were knowledgeable on environmental matters in the division of power. The intervening years had made clear that environmental impacts were always a factor, frequently the controlling one, in every electric utility function. The environmental division we created was headed by Dennis Rapp, who had a lot of related experience in Washington. If the PSC only stressed environmental considerations, there might be little economic life. It was necessary to make choices, to adjust programs so that the production of economic goods was accommodated in a mix that provided the greatest possible public benefit in the long run. I was sometimes surprised, or even disappointed, at the extent to which our own staff made common cause with the state's Environmental Conservation Department. The concept of the best mix for the public good was foreign to them. They had a sense of mission and made poor compromisers.[4]

The other addition to the organization was an office of economics. The economics staff at the FPC had been useful in formalizing the inclusion of economic factors in the decisional matrix. Although I had become wary of the ease with which economists disposed of multi-dimensional problems by reference to economic dogma, their input had been helpful. I contacted several individuals about heading the FPC's economics office, including Alfred Kahn (his specialty was regulated industries), but he declined. Alvin Kaufman, a practicing economist in Washington, applied for the job. He was one of the numerous scholars who had finished all the courses for a Ph.D. degree at a good university. For one reason or another, however, he had not received the degree for lack of a dissertation. I preferred a man with all possible diplomas, ribbons, and decorations. Nonetheless, Kaufman impressed me favorably and got the job. I was happy afterward with the choice. He was sound, had an excellent background, could hold his own in contests with other economists, and had good judgment and sense, qualities not always found even in economists of high reputation. He was also a prodigious worker. It was he who provided the initial spark for the creation of an economic research institute within the framework of the National Association of Regulatory Utility Commissioners (NARUC). His suggestion was excellent, and I pushed it along as well as I could from my position as a member of the NARUC Executive Committee. The eventual outcome was the National Regulatory Research Institute (NRRI).[5]

Among the new commissioners appointed during my tenure was Carmel Carrington Marr. Marr was brought to my attention by Ersa Poston, the head of the New York State Personnel Department. I asked her help in recruiting an

African American for the commission, and she suggested Marr. She proved to be a superb choice: gentle, soft-spoken, a good listener, of sound judgment, and firm in her opinions when they were reached. She became easily the most popular of the commissioners. Marr had an African American constituency, and she was careful not to alienate it. What this meant, as a practical matter, was that she was at the low end of the zone of reasonableness in rate cases affecting New York City consumers.[6]

For some thirteen months [following Swidler's appointment] the PSC exercised regulatory power over railroad, truck, and bus transportation, after which the Transportation Act turned over jurisdiction to the Department of Transportation. We were relieved to be rid of it. The buses and local traction companies were losing money and required subsidies. We had no money for subsidies. On the other hand, intrastate railroad passenger service did not amount to much. Utility regulation is designed for profitable industries, not for industries overwhelmed by competitive technologies. Bus service was still important, but it could not pay its way, at least in intrastate service. The transfer was as it should be.[7]

Truck transportation was different. Most of the lines made money, though only by dint of securing rate increases promptly. If the truck drivers won a wage increase, the companies required rate relief in a matter of weeks. It was difficult to check their accounts since there was no uniform system of accounting. Ratemaking by the PSC was therefore a seat-of-the-pants exercise. It was clear to me that regulation of motor carriers was not working. I was happy that I had no responsibility to propose changes.

CHAPTER 27

Regulation in an Inflationary Environment

The changes made to the commission's procedures conformed to the innovations that had proved effective while I was at the FPC. Meetings were held in the presence of the senior staff, where any dereliction by a commissioner would become known. ALJs prepared decisions independently, and any party could receive commission review by filing exceptions to the ALJ report. A group of lawyers with exceptional analytical and writing abilities was recruited to examine the record and briefs in each case. They also prepared drafts of opinions in accordance with the commission's instructions. The ALJ group was strengthened by some imports from Washington. The ALJs were happy in their new freedom within a structured process; they took great pride in their work and performed as well as one could wish.

The New York Telephone Rate Case

At my first commission meeting, the principal item on the agenda was a New York Telephone Company rate case, overripe for decision. The amount applied for was $175 million, but the company had not received a general rate increase since 1957. New York Telephone served New York City and much of the rest of the state. Its revenues and invested capital were more than ten times greater than the combined figures for the other fifty-six telephone companies in the state, including the Rochester Telephone Company.

Having given up hope of a reasonable rate award, New York Telephone had refused to invest the huge sums, at least a billion dollars a year for the next decade, that were needed to restore high quality service. The reason was not an inability to raise money. The company, a subsidiary of AT&T, could have thrown more money into the pit, but it was disinclined to do so. The result was the virtual breakdown of telephone service in New York City—and low quality service elsewhere in New York. If New York Telephone and AT&T had used poor service to gain the attention of the commission, the ploy did not work on my predecessor. He did not deny rate increases; he merely delayed action indefinitely. At first, I was disappointed that he had not disposed of this case. I realized, however, that perhaps unwittingly he had done me a favor. The service problem arose out of the money problem, and New York Telephone needed a signal of fair treatment ahead.

The situation was much like the one I had met in Washington when the Tennessee Gas Transmission Company refused to expand its service in the New England area. In both cases I might well have considered that the company was engaged in a test of strength. If so, I could have recommended to the commission that we order the improved service and await compliance before taking up the question of rate relief. We might have won the contests in the courts, but there were several difficulties with this approach. In the first place, it wasn't fair. In both cases, the companies could make a strong argument on constitutional due process grounds. Doubtless, the litigation would have been appealed to the U.S. Supreme Court, with a final decision years away. In the meantime, consumers would suffer. I could imagine that if I were on the other side, I would probably follow the course of the utility executives. The companies were not our enemies, and if we lived up to our obligations under the regulatory compact, they would improve service far more quickly, and at lower cost, than if under compulsion. The telephone companies knew that if they did not comply with alacrity after having received reasonable rate consideration, they might not receive kind treatment in the future. They might even lose their service franchises.[1]

Commissioner Edward Larkin submitted a proposal in February permitting an allowance of $137 million. It was treated as a recommendation by an ALJ, subject to exceptions and review after briefs and oral argument before the commission. As modified after rehearing, the commission allowed a $105 million rate increase.[2]

Innovations in Telephone Regulation

The commission not only responded to rate case filings, but it also introduced several new regulatory features to provide a check-and-warning system on quality of service. For one thing, the system of processing complaints was strengthened. In 1969 the commission handled 16,399 complaints. In 1970 the number rose to 40,207, a 145 percent increase. These data do not reflect the full magnitude of the crisis since the commission did not have the people and facilities to handle all the complaints. By the end of 1970, there were twenty people on the complaints staff, some in New York City and others in Albany. The two groups were interconnected in such a way that if the people in New York City were overloaded, the calls were shifted to Albany, and vice-versa. A facsimile machine transmitted written reports from both offices to New York Telephone's offices. The average time required to handle a complaint was reduced from thirty days to eight. Computer analysis made it possible to identify the trouble spots and expedite action by the company. Computer technology also provided a ready way to follow up on New York Telephone's efforts and to stimulate improvements.

In addition to linking rate awards with quality of service, the commission introduced direct and automatic financial incentives, including credits to consumers for service interruptions controllable by the company and not corrected within twenty-four hours. The credit was two-thirtieths of the monthly tariff rate for each twenty-four-hour interruption after the first twenty-four hours; that is, twice the tariff rate per day. This requirement was applied to all telephone companies and to Western Union.[3]

The commission concluded that full assurance of high quality service would require systematic inspections. We had inherited a telephone investigation staff of five people. This was nowhere near the number we needed. The staff was expanded to eighteen in the first year. Every telephone exchange in the state would now receive an annual visit. In this way the commission would be made aware of problems and could take steps to correct them before a breakdown occurred. These visits also permitted the commission to become involved in system planning at an early stage.

The Basic Service Index Plan (the first of its kind in the country) provided for a $1.50 per month rebate to each telephone subscriber in any telephone exchange for chronically poor service. The quality of service was defined in precise terms: the seconds required for a dial tone, for operator response, or for call completion. The highest possible score was 165. The commission determined that a three-month rolling overage of 120.94 or lower warranted a rebate. These rebates would not be treated as an operating expense for rate purposes. Our guess was that as many as two million subscribers in the downstate area would be eligible for refunds. Within ten months service improved to the extent that only 100,000 refunds were made for a total cost of $200,000. In 1973 service had improved to the point that no refunds were required.[4]

In the interest of monitoring New York Telephone's quality improvement program, the commission ordered the company to make monthly progress reports. One improvement gave me special satisfaction. At the beginning of 1970, about 20 percent of all the coin phones in New York City were out of service at any given time. The situation was particularly bad in Manhattan and Brooklyn. Pay phones are the principal resource of the poor and of strangers. The company said it was impossible to do better because of vandalism and theft of coins and tokens. The commission said that these excuses were inadequate and that the situation required greater effort. We asked whether the pay phones and the enclosing structure could be made more resistant to vandalism and theft. The company took the problem back to Bell Labs and to Western Electric, the supply arm of the AT&T system. Soon thereafter the cords were steel sheathed, the coin boxes were strengthened, and most of the phones were set on open pedestals where vandals and thieves could not work undetected. The company also increased its repair forces. By the end of the year, the outage rate for pay phones had been halved.[5]

A proceeding in which I developed a particular interest involved hotel, motel, and hospital surcharges for telephone service. Many of these institutions were charging varying rates for outgoing calls: flat charges as high as $1.50 a day for telephone availability (whether or not used) and extra charges for intrastate long distance calls. The commission undertook an investigation of the surcharges. I was soon visited by a lobbyist for the hotel and motel groups. He told me of the large telephone investments and operating expenses that were required for both incoming and outgoing calls. I said I could readily understand that the costs were substantial, but that a good case could be made that they were as much a part of the essential costs of operating hotels and motels as the costs of chambermaids and security personnel. He then proceeded to explain how the separate charges could and should be justified by the commission. I told him I had been hired to protect consumers, not to find ways to justify rate increases. He was shocked—shocked!—by my failure to see things his client's way.

The enforcement of commission orders on hotels and motels was a ticklish question. The commission was unable to check on the thousands of installations that were involved. Instead, we ordered New York Telephone to refuse to serve recalcitrant hostelries. The company seemed entirely willing to comply, probably because swelling the costs of its service by third party surcharges was not a good way to encourage growth in its volume of business. At any rate, when the staff proposed, and the commission approved, small charges for outgoing calls (and none for availability), there was almost universal compliance. Out of 706 hotels that were canvassed, only one failed to heed the order when reminded.[6]

The battle was won, but it was a Pyrrhic victory. The hotel industry did not want to take responsibility for an increase in room rates if it could help it. In the inflationary atmosphere of the times, though, another dollar or two in the room rate did not prove to be a significant problem for most hotels. The only victory the PSC could claim was a contribution to greater truth in labeling.

Electric Power Problems

In New York, growth [in electric power] had increased by 60 percent in the 1960s. Everyone knew that this pace could not continue indefinitely. The utilities, however, had to commit themselves to reliance on the growth curve of the past in planning for the future. The PSC's annual report for 1970 pointed out that even though the number of electricity customers in the state was increasing at the rate of only 1 percent, electricity usage was growing at a rate of 6 percent.

At the beginning of 1970, the supply picture was tight, but not alarming. There was apparently enough for summer usage without lowering the quality of supply during outages. When Con Ed lost its 265-MW Indian Point nuclear station in May, the situation began to deteriorate. Then, on June 11, Con Ed was compelled to reduce its output by 3 percent. While not serious in itself (a reduction of 3 percent will dim lighting slightly, distort television images a little, and slow motors), it was a precursor of worse to come. The situation became really critical on July 21 when Con Ed lost its largest unit, Ravenswood 3, which itself represented 11 percent of Con Ed's system capacity. An 8 percent reduction is at the limit of feasibility. Throughout the whole summer, and continuing into the fall, there was great uncertainty. I began to make weekly reports to the governor on power supply problems in the state. They were released to the press and received attention until the situation stabilized.[7]

Between the initial brownout on June 11 and the end of 1970, utilities were forced to reduce voltage by 3, 5, or 8 percent. The utilities could not even relax at the beginning of fall. An atypical heat wave that hit the state on September 22 caused the greatest power crisis since the 1965 blackout. Each of the state's seven major utilities reduced voltages. Con Ed, for the first time in recent history, was forced to impose a complete blackout on a part of its system in order to preserve service for the rest. Orange & Rockland, Long Island Lighting, and Central Hudson were also forced into partial blackouts. The interruptions were short, but the uncertainties of supply were a burden on consumers and a constant source of tension for the utilities. I had been with the commission so short a time that no one thought to implicate me as a contributor to the shortage problem.

The possibility of shortages, brownouts, and blackouts during the hottest days of the year was not without its lessons. It illustrated the willingness and ability of consumers to cut back their use voluntarily when called upon to do so. It indicated that the utilities had profited from the lesson of the 1965 blackout: They had to cut off parts of their systems as necessary in order to prevent system collapse. It proved the value of the existing transmission connections in bringing in power from distant sources, including the Ontario and Quebec systems. It also demonstrated the need to strengthen the transmission systems and to cure bottlenecks. Above all, it made a case for coordinated, long-term power planning.

The Legislative Program on Power Supply and the Environment

In the early months of 1970, I spent much of my time writing drafts of proposed laws to assure that system planning took full account of environmental impacts. I negotiated with political leaders and representatives of environmental organizations to secure their support of this legislation. The Senate was generally receptive to my proposals. The Senate leader, John Anderson, was a friend and supporter of Governor Rockefeller. He was predisposed to accept the recommendations, though he exercised independent judgment as well.[8]

The situation in the Assembly was much different. It was more likely to focus on parochial interests. The Assembly leader, Stanley Steinberg, was a friendly man, anxious to help anyone who asked. He took his responsibilities seriously. As a professional politician, he reflected the views of his constituents and colleagues. Steinberg was not inclined to serve as a middleman. Instead, he referred me to an environmental lawyer, Irving Like, a leader on conservation issues in the New York City area. Steinberg told me that he would accept anything Like would agree to.[9]

I spent many hours in negotiations with Like. After a few months, it appeared that we would be unable to agree in time for action on a comprehensive bill for the 1970 legislative session. We concurred on the transmission-line features of a draft bill, but not the generating-plant provisions. To cinch this partial progress, we made the transmission-line features the subject of a separate bill that became law on July 1, 1970. The law required electric utilities to obtain a certificate from the PSC authorizing the construction of high voltage transmission lines (natural gas transmission lines, although involving somewhat different environmental and safety problems, were also covered by the statute). In the interest of procedural efficiency, the law established a single-step process, in which both public need and environmental compatibility were to be evaluated during the proceeding. The utility was to advertise its proposed actions; public hearings were required; and all affected parties and interested agencies were invited to participate. The PSC extended an explicit invitation to the state environmental agency, which participated in every proceeding under the statute.[10]

The PSC's Significance

What the commission brought to its job that was unusual, if not unique, was a willingness to recognize that costs were skyrocketing, and that these costs might point to unpopular rate increases. Rate policies, as a nationwide matter, provided no opportunity for the utilities to set up reserves in good times that would permit leveling rates in lean times. At best, as a group, they received a minimum return. The PSC did not, however, neglect the interests of consumers. To soften the impact of higher costs on consumers, we developed options for minimal service at reduced rates.

PART IX

Final Years, 1974–1997

Swidler concludes by summarizing his views on regulation. He indicates that regulators should be highly qualified individuals willing to cooperate with their counterparts in other public agencies. Furthermore, he again emphasizes the importance of balancing the needs of consumers and producers. The regulator should encourage producers (via incentives) to reduce costs so that the end result will be *lower* rates. Inflationary periods, however, might necessitate *higher* rates, which will inevitably generate consumer protests. According to Swidler, the regulator should stand firm against these outcries—a piece of advice that likely reflected the pressures he himself had experienced during his tenure at the PSC.

Swidler expresses a degree of ambiguity regarding the related issue of competition. He did not accept the contention that utilities should directly compete with one another in an unfettered market. Yet he also rejected the assertion that regulated utilities are immune to the healthy competitive pressures that keep rates low and the quality of service high. To Swidler, these views might not have been as contradictory as they seem. The distinction he appears to have been making pertained to the circumstances in which competition occurred. In a regulated environment, competition would yield positive results; in an unregulated environment, it would produce negative ones. Whether right or wrong, such a distinction was typical of Swidler's efforts, throughout his life, to recognize that policymaking was a complex undertaking that defied simple solutions.

After leaving the PSC, Swidler helped establish, and briefly headed (1974–75), the Institute of Public Policy Alternatives at the State University of New York at Albany. The Swidlers then returned to the Washington, D.C., area, where he practiced law and Gretchen developed a career in real estate. In the firm that bore his name (Swidler & Berlin), Swidler enjoyed playing the role of elder statesmen, talking to the younger attorneys, and charming the staff in a courtly manner. He performed necessary legal work and brought in new business, but was always more engaged in public issues than in the relatively mundane tasks of recruiting big clients and serving their interests. He gave speeches, presented testimony, and sat on the advisory councils of the Electric Power Research Institute, the National Academy of Public Administration, the Gas Research Institute, and the board of directors of the National Regulatory Research Institute. From these and other positions, Swidler offered his opinions on the direction and future of U.S. energy policy. While he did not directly address the rise of political conservatism in the 1980s, his comments on other topics make it clear that he was troubled by the newly enhanced influence of the private sector in national and state policymaking. Those comments—like this memoir, which he was still working on at the time of his death in 1997—were not occasions for polemics and diatribes; rather, they were Swidler's attempts to present reflective and instructive meditations on the nature of public and private power. Judged by this standard, Swidler was successful.

CHAPTER 28

A Short Apologia for Utility Regulation

This is a fitting place to set out my view on the policies that should govern regulation of the investor-owned utilities and of the electric power industry in particular.

Regulation

The primary purpose of regulation is to insure that consumers will receive adequate and reliable service at reasonable rates. The obligation to serve includes the obligation to build for the future, even in the face of uncertainty. Since shortages are intolerable, a margin of available capacity over probable needs is always desirable, perhaps even essential. This margin can be achieved by more efficient use of electricity, by managing the load to reduce peaks, by building new capacity in both generation and transmission systems, or by a combination of the three.

Utility regulation is cost based, and rates depend on costs. Regulators should give priority to helping utilities cut costs; this should provide a basis for reducing rates. Of course, cost reductions need to be based on long-term considerations. Skimping on quality or reliability of service is untenable. Inevitably, costs will be influenced by the increasing need to protect the air, land, surface waters, and underground water supply.

Since most consumers focus on their present utility bills, the likely effect of current rates on future prices holds little interest for them. The press encourages this shortsighted view. Prevailing short-term pressures tend to tilt regulation toward denying adequate rates under inflationary conditions. Regulators should realize that if they apply long-term criteria in doing their jobs, regulation will not be a lifetime occupation.

Rate and cost comparisons should be based on internalization of social costs, such as those associated with the environment and government services. Competition based on insufficient environmental standards or inadequate transfer costs does not serve the public interest. Functionally, the electric utilities are primarily converters of fossil fuels into electric energy, which involves releasing vast amounts of pollutants and heat into the atmosphere and surface waters. Wastes from generating plants may contaminate underground aquifers. Environmental protection has become and should remain a principal responsibility of utility regulators.

Regulatory agencies should be well staffed, and only by those motivated to do an effective job. State and federal agencies must collaborate to maximize the effectiveness of regulation, especially in avoiding conflict in situations of mutual concern. Utilities, with the support of regulators, should be in the forefront of developing and adopting technological improvements. Close-working relations with utilities should not mean being co-opted by them. The rules against ex parte communications should be strictly enforced, and social relationships between regulators and officials of regulated companies should be avoided. The same rules should be applied to the relations between regulatory agencies and public systems.

Competition

A controversy has been raging for decades on the future of utility regulation. The debate has focused on the extent to which competition should replace regulation in the electric power field. The common argument runs something like this: Instead of minimizing expenses, cost-plus regulation actually encourages excess investment and high costs. The cure is said to be competition.

The introduction of some elements of competition may be desirable. Regulated utilities, however, are different in many ways from companies that are part of the free market system, a system in which firms can go out of business or move to other locations. Utilities are unique in being allowed to use streets and highways for their facilities, and in their ability to exercise the right of eminent domain. Utilities are accorded these privileges because without them they could not provide their services to the public. The economics of the industry are based on the integration of systems. This requires cooperation among companies in planning and operating generating plants and transmission lines. The benefits that flow from cooperation would not survive head-to-head competition.

The claim that regulated utilities lack incentives to do a good job is grossly overstated. This is not what I have observed in the past, although the situation may change as certain "reforms" take hold. There are many existing pressures that encourage—and have indeed produced—high levels of performance. For example, to do well, a utility must enjoy the respect and confidence of the people in its service area. Consumers can always look in a different direction, municipalization primarily, if they lose confidence in their electricity provider.

In the last couple of decades, new pressures for performance have been instituted. The state public utility commissions have adopted the practice of requiring independent performance reviews by outside firms of engineers, accountants, and economists. Typically, these reviews have given the utilities high marks, coupled with suggestions for peripheral improvements, most of which were accepted and promptly placed in effect. Published rankings of utilities, from best to worst, have also been helpful. Companies with rates at the low end are subject to pressures to improve their performance or to provide public justification of costs and rates.

The impression given by critics that regulated utilities are immune to competitive pressures is also an exaggeration. They operate in glass houses. All their books are open to the public. They cannot change prices without public hearings, at which consumer agencies are well represented. There is always competition, or the threat of competition, especially at the periphery of the service area. This gives wholesale customers a choice of whether to buy from one source or another. Furthermore, there are the unofficial evaluations of bond rating agencies and financial analysts, as well as the messages transmitted by company earnings, dividends, and stock prices. In the aggregate, the pressures to do a good job are as strong as in the so-called competitive sector, where the public and stockholders have little on which to base an appraisal of relative performance.

Fortunately for consumers, the behavior of utility officials has differed from that of their peers in various competitive industries. Utilities perform a public service. They cannot take a short view because they are tied to the fortunes of a business with a long history and presumably a long future. They cannot move to a more favorable location and they cannot curtail service when business is bad. They cannot choose their customers, or confine themselves to the most profitable segments of the market. They must serve all who apply, without discrimination. They lack flexibility to increase rates to cover rising costs. Even when earnings do not warrant new investments, or when the cost of money is abnormally high, they must raise new capital for future needs. They work together to reduce costs by forming pools, building joint plants, maximizing the economics of their combined systems, and by supporting joint research. They are good citizens who live where they work. They have a profound interest in the viability of the public and private institutions upon which they and their customers must rely.

What needs changing the most is the system of incentives for improving performance. We must introduce a better system of reconciling risks and rewards. The stick is not an adequate spur to optimum performance, yet it has been the principal regulatory tool.

It is as bad for consumers when rates are set too low as it is when they are set too high. If rates are set so low as to impair a utility's credit, it cannot raise money on favorable terms. The increased cost of money becomes a part of the cost of service. Rates will rise accordingly. The consumer will be faced with high prices for poor performance. A utility with impaired credit will encounter other costs. Maintenance might have to be deferred. Managements might lose their most talented people. The utility might be unable to make advantageous purchases of fuel and equipment. It might also hesitate to risk adoption of even the most promising new technologies. The competition among regulators should not be to determine which among them can come closest to driving a utility to the edge of bankruptcy. Instead, they should focus on finding a sensible solution that provides companies with earnings high enough to enable them to hold costs down for the long term.

The basic values of utility managers, their sense of responsibility and accountability, would erode if the utilities were thrown into direct competition for each other's customers. This would be a high price to pay for accepting the contentions of the single-minded independent power entrepreneurs. They have no responsibility for power supply or for community welfare, except for specific contractual obligations. Free competition, in their usage, includes seeking and receiving many statutory benefits. Their only argument is to trumpet the magic of competition, regardless of circumstances, while they strive to win special legislative favors. There might be some benefit from extending and intensifying this one-way competition, but there is also much to lose.[1]

Notes

Editor's Introduction

1. Joseph Swidler, "An Unthinkably Horrible Situation," *Public Utilities Fortnightly* 128 (Sept. 15, 1991): 14–18, 40, quote at 16; Jordan A. Schwarz, *The New Dealers: Power Politics in the Age of Roosevelt* (New York: Random House, 1993), quote at 235.

2. The Public Utility Regulatory Policies Act of 1978 (PURPA) is located at 92 Stat. 3117. Swidler discusses the concept of a regulatory compact in chap. 27. For contemporary discussions about the electric power industry, see "Record Rate Hike Set in California," *New York Times* (hereafter *NYT*), Mar. 28, 2001, A1, A16; "Trying to Shore Up Utilities, California Plans Rate Increase," ibid., Dec. 22, 2000, A16; "A Dwindling Faith in Deregulation," ibid., Sept. 15, 2000, C1; "In Reverse: California Acts to Cap Some Electric Bills," ibid., Aug. 22, 2000, A12; "Energy: The Power to Choose," ibid., Oct. 19, 1999, EN1-10; "Mutiny on the Meter," *Washington Post,* Dec. 3, 2000, H1; "Pepco to Sell Power Plants," ibid., Feb. 4, 1999, A1, A10.

3. Harry R. Booth (1902–1974), a graduate of the University of Chicago, was an attorney who specialized in utilities law. He was a legal assistant to David Lilienthal (1927–31), counsel for the Illinois Commerce Commission (1933–40), chief of the Federal Communications Commission's rate section (1941), and chief of the utilities branch of the Office of Price Administration (1942–46). For a representative sample of Lilienthal's personal reflections, see David E. Lilienthal, *The Journals of David E. Lilienthal: The TVA Years,* vol. 1 (New York: Harper, 1964).

Part I. Education, Private Practice, and the Pull of Public Service

1. The literature on the rise of utilities and holding companies, as well as their regulation, is vast. Useful studies include Mark H. Rose, *Cities of Light and Heat: Domesticating Gas and Electricity in Urban America* (University Park: Pennsylvania State Univ. Press, 1995); Harold L. Platt, *The Electric City: Energy and the Growth of the Chicago Area, 1880–1930* (Chicago: Univ. of Chicago Press, 1991); Thomas K. McCraw, *Prophets of Regulation: Charles Francis Adams, Louis D. Brandeis, James M. Landis, Alfred E. Kahn* (Cambridge: Harvard Univ. Press, 1984); Phillip J. Funigiello, *Toward a National Power Policy: The New Deal and the Electric Utility Industry, 1933–1941* (Pittsburgh: Univ. of Pittsburgh Press, 1973); Ellis W. Hawley, *The New Deal and the Problem of Monopoly* (Princeton: Princeton Univ. Press, 1966). Also helpful is Forrest McDonald, *Insull* (Chicago: Univ. of Chicago Press, 1962). A good analysis from the 1930s is James C. Bonbright and Gardiner C. Means, *The Holding Company: Its Public Significance and Its Regulation* (New York: McGraw-Hill, 1932).

2. The TVA's enabling legislation is located at 48 Stat. 58. In 1935, Congress passed an amendment authorizing the TVA to regulate the resale (retail) rates charged by

distributors for its electricity. As Swidler and other sources show, however, the TVA was specifying retail rates from almost the beginning of its operations. See 49 Stat. 1076; C. Herman Pritchett, *The Tennessee Valley Authority: A Study in Public Administration* (Chapel Hill: Univ. of North Carolina Press, 1943), 83. Swidler's salary at the Interior Department was $3,500. By Depression-era standards, this was a comfortable income; it paled, though, in comparison to private-sector salaries. In the 1920s, Harvard-trained lawyers were earning an average of $10,000 a year. At the TVA, Swidler's annual salary would increase to $6,000 by 1936. See TVA, *Annual Report for the Tennessee Valley Authority for the Year Ended June 30, 1936* (Washington, D.C., 1936), 368.

CHAPTER 1. EARLY YEARS

1. Vitebsk (sometimes spelled "Vicebsk") is a city and province in northern Belarus. Swidler was born on January 28, 1907. His two sisters were Ethel Swidler Glickstein (1906–1988) and Bertha Swidler Miller Goldstein (1909–). His parents were Dora Cromer (born in Vitebsk) and Abraham Swidler (born in Myotl, Russia). When Swidler was born, his father was thirty and his mother was twenty-eight.

2. Nicola Sacco (1891–1927) and Bartolomeo Vanzetti (1888–1927), Italian-born anarchists, were arrested on charges of robbery and murder on May 5, 1920, convicted on July 14, 1921, and executed on August 23, 1927.

3. Willis Van Devanter (1859–1941; associate justice, 1910–37); James Clark McReynolds (1862–1946; associate justice, 1914–41); George Sutherland (1861–1942; associate justice, 1922–38); Charles Evans Hughes (1862–1948; associate justice, 1910–16, chief justice, 1930–41); Harlan Fiske Stone (1872–1946; associate justice, 1925–41, chief justice, 1941–46); Oliver Wendell Holmes (1841–1932; associate justice, 1902–32); Louis Dembitz Brandeis (1856–1941; associate justice, 1916–39); Benjamin Nathan Cardozo (1870–1938; associate justice, 1932–38).

4. Arthur Goldberg (1908–1990) was U.S. secretary of labor (1961–62) and associate justice of the Supreme Court (1962–65). Philip Glick (1905–), like Swidler, received his undergraduate (1928) and law degree (1930) from the University of Chicago. He served as general counsel for the Federal Subsistence Homesteads Corporation (1933–34), assistant solicitor in the U.S. Department of Agriculture (1934–42), solicitor (1942–44) and deputy director (1945–46) of the War Relocation Authority, general counsel for the Federal Public Housing Authority (1946–48), and counsel for the Institute of Inter-American Affairs (1948–53). David Krooth (1907–1978), a fellow University of Chicago classmate, was counsel for the Federal Emergency Administration (1933–37), associate general counsel for the Federal Public Housing Authority (1937–42), general counsel for the Federal Public Housing Authority and Defense Homes Corporation (1942–46), and general counsel for the National Housing Agency (1946).

5. Created in 1916 by the National Defense Act (39 Stat. 213), the Reserve Officers Training Corps (ROTC) prepares high-school and college students to become officers in the U.S. armed forces.

6. The tenures of the University of Chicago faculty noted by Swidler were: James Parker Hall, 1902–28; Edward Wilcox Hinton, 1913–36; George Gleason Bogert,

1926–50; Ernst Freund, 1902–33; Floyd Russell Mechem, 1903–29; Ernst Wilfred Puttkammer, 1922–56; Harry Augustus Bigelow, 1904–40; Frederic Campbell Woodward, 1917–30 (vice president of the university, 1926–40).

7. Arthur Harold Kent (1894–1960) taught law at the University of Chicago from 1927 to 1934. He subsequently became the assistant chief counsel and acting chief counsel for the Bureau of Internal Revenue (1934–36) and assistant general counsel for the U.S. Treasury Department (1936–38).

CHAPTER 2. LAW PRACTICE IN CHICAGO

1. Since the interest derived from municipal bonds is exempt from federal taxation, municipalities normally pay less interest to their bondholders than do private, for-profit companies. What the bondholders forfeit in reduced interest is compensated by a lower tax bill. The ability to borrow money more cheaply (than the private sector) gives publicly owned utilities an economic advantage. The Public Ownership League of America was organized in 1914 to promote public ownership of utilities, natural resources, and, later, railroads. With national headquarters in Chicago, the League's supporters included Jane Addams (1860–1935) and Frederic Howe (1867–1940). Albert May Todd (1850–1931), a Democrat and a U.S. representative from Michigan (1897–99), was president of the League until 1922. The League's national journal, *Public Ownership of Public Utilities*, ceased publication in 1950. A review of its initial activities can be found in Public Ownership League of America, *Proceedings: Public Ownership Conference, November 15–17, 1919* (Chicago: Public Ownership League of America, 1919).

2. Donald Richberg (1881–1960) was a prominent Chicago attorney and one of the founders of the National Progressive League (1911). He gained widespread attention in the 1920s as a labor lawyer, and later served as chief counsel for the National Recovery Administration (1933–35). Though branded as a "radical" by some conservatives, Richberg adhered to the tenets of mainstream liberalism. It is likely that Swidler's admiration for Richberg was typical of those who supported working-class interests without calling for significant revisions in America's capitalist framework. David E. Lilienthal (1899–1981) was a member of the Wisconsin Public Service Commission (1931–33), director (1933–41) and chairman (1941–46) of the TVA, chairman of the U.S. Atomic Energy Commission (1946–50), and chairman of the board of the Development and Resources Corporation (1955–78). For the most recent analysis of Lilienthal's life and career, see Steven M. Neuse, *David E. Lilienthal: The Journey of an American Liberal* (Knoxville: Univ. of Tennessee Press, 1996).

3. Edward W. Morehouse (1896–1974) taught economics at Northwestern University (1925–31), was an editor of the *Journal of Land & Public Utility Economics* (1925–34), chief economist for the Wisconsin Public Service Commission (1931–40), a consultant to the TVA during the 1930s and 1940s, and vice-president of the General Public Utilities Corporation (1945–61). Swidler wrote two articles for the *Journal of Land & Public Utility Economics*: "Rate Regulation in California: Some Significant Problems Raised by the California Railroad Commission," 7 (Feb. 1931): 1–12, and "Utility Rate-Making in the Depression," 8 (Nov. 1932): 347–51. In both

articles, Swidler grappled with the complex financial and legal issues confronting regulators and courts during the Depression. While Morehouse's influence on these articles is not entirely clear, both men believed that public oversight of private utilities was necessary.

4. Philip La Follette (1897–1965) served as a Republican (1931–33) and Progressive (1935–39) governor of Wisconsin. His father, Robert M. La Follette (1855–1925), was an influential reformer ("Fighting Bob"), a Republican representative from Wisconsin (1885–91), governor of Wisconsin (1901–6), and a U.S. senator from Wisconsin (1906–25). Philip's brother, Robert M. La Follette Jr. (1895–1953), was a U.S. senator from Wisconsin who served as a Republican (1925–35) and a Progressive (1935–47).

5. Paul H. Douglas (1892–1976), a Democrat, was an economics professor at the University of Chicago (1920–49), an adviser to the National Recovery Administration, and a U.S. senator from Illinois (1948–66). Harold Ickes (1874–1952) was best known as a Republican and Progressive political reformer prior to becoming U.S. secretary of the interior in the Roosevelt and Truman administrations (1933–46). He also headed the Public Works Administration (1933–39). See Jeanne Nienaber Clarke's excellent study, *Roosevelt's Warrior: Harold L. Ickes and the New Deal* (Baltimore: Johns Hopkins Univ. Press, 1996). There is little surviving information about the Utility Consumers and Investors League. Formed in April of 1932, the League argued that utility rates in the Chicago metropolitan area were far higher than the market justified. The League called for changes in the administrative structure of the Illinois Commerce Commission (ICC), as well as judicious experimentation with public ownership. The League's members included Paul Douglas, Donald Richberg, and Charles Merriam (1874–1953), a University of Chicago political scientist. Swidler's partner, Harry Booth, was retained as the League's counsel. A statement of the League's concerns and recommendations can be found in a report entitled "To the Consumers of and Investors in the Public Utilities of Illinois," Apr. 7, 1932, box 57, David Lilienthal Papers, Seeley G. Mudd Manuscript Library, Princeton University.

6. Samuel Insull (1859–1938) was born in England and immigrated to America in 1881 to be Thomas Edison's private secretary. Beginning in the 1890s, and continuing over the next four decades, Insull used holding companies to create a vast utilities empire. When that empire suddenly collapsed in 1932, many contemporaries blamed Insull himself; he was eventually acquitted of a number of charges, including fraud and embezzlement.

7. In 1929, an enormous deposit of natural gas, the so-called Amarillo natural gas field, was discovered in the panhandle of Texas. It was approximately one million acres. Several investors, including Insull, constructed a twenty-four-inch pipeline from Fritch, Texas, to Joliet, Illinois. Insull then built two transmission lines from Joliet to the three Chicago gas distributors—Peoples Gas, Public Service Company of Northern Illinois, and Western United Gas and Electric Company. Insull Utility Investments owned large shares of stock in all three companies. The mixture of artificial and natural gas was introduced on Oct. 16, 1931. See ICC, *Opinions and Orders* (Springfield, 1931), 11:524; ICC, *Opinions and Orders* (Springfield, 1932), 12:1464–65.

8. Swidler's discussion of this case requires clarification. During the initial hearings (June 25–26 and July 13–15, 1931), Swidler and Booth represented the Amalgamated Improvement Associations, not the Utility Consumers and Investors League. The latter group opposed a Peoples Gas bond issue in June of 1932. See "Stock Buyers' League Attacks Utility Finance," *NYT,* June 10, 1932, 29. Booth and Swidler attempted to persuade the ICC to force Peoples Gas to show cause why the gas consumers of Chicago should not benefit from the cheaper natural gas to be piped from Texas. As Swidler notes, the ICC did permit Peoples Gas to upgrade the energy level of its artificial gas. The ICC, however, ultimately required that Peoples Gas implement a 6.5 percent rate *reduction,* effective Oct. 15, 1931. At the time, and in retrospect, Swidler and others were likely disinclined to view the ICC's decision as one favorable to consumers. There was significant anecdotal evidence that gas bills increased, not decreased, after rate reductions were implemented. For example, on Oct. 19, 1931, Peoples Gas received six thousand complaints concerning higher gas bills. The Utility Consumers and Investors League later speculated that the inability of existing gas appliances to use the newly enriched gas in an efficient manner might have been the reason for increased costs. Furthermore, the League claimed that Peoples Gas accrued unjust profits. The company only paid 2 cents per thousand cubic feet for the Texas gas it purchased outright. After taking into account the costs associated with piping this gas to Illinois (approximately 18 to 45 cents per one thousand cubic feet), Peoples Gas still made large profits (even a 6.5 percent reduction in rates would have meant that the company was charging consumers $1.26 per one thousand cubic feet of gas). See "Insull to Halt Gas Rate Cut, Foes Concede," *Chicago Herald and Examiner* (hereafter *CHE*), July 20, 1932, 4; ICC, *Opinions and Orders,* 11:523–26; "Open Hearings on Plea to Mix Artificial Gas: City Request for Delay is Overruled," *Chicago Daily Tribune* (hereafter *CDT*), June 26, 1931, 8; "Demand Rates for New Gas Justified," ibid., July 14, 1931, 7; "End Testimony on Petition for Mixture of Gas: ICC Ponders Plea of Company," ibid., July 16, 1931, 11; "Gas Rate Hearings May Start Today: May Last Six Months," ibid., Sept. 30, 1931, 8; "Order 6 Percent Cut in Domestic Gas Rates Here: Commission Ruling Also Affects Suburbs," ibid., Oct. 2, 1931, 3; "Stock Buyers' League Attacks Utility Finance," ibid., June 10, 1932, 29; "University Aids City's Fight for Lower Gas Rate: Finds Its Bill Increased 50 Percent by New Fuel," ibid., June 1, 1932, 19; "To the Consumers of and Investors in the Public Utilities of Illinois," 3.

9. Both Swidler and Booth were cited for contempt on July 21, 1931; they apologized to the ICC on July 28, 1931. See "Insull to Halt Gas Rate Cut, Foes Concede," 4; "Lawyers Apologize to Commerce Body," *CHE,* July 29, 1931, 4 (copies of both articles are located in box 2, subseries 2, folder 2-14, Harry R. Booth Papers, Chicago Historical Society). Lilienthal wrote to Charles W. Hadley of the ICC, indicating that he thought Swidler and Booth had acted "unfairly" and "improperly." He hoped, however, that the ICC would drop the contempt charges. This request seems to have been motivated by Lilienthal's fear that additional publicity could undermine his publishing venture with the Commerce Clearing House. "It might destroy," Lilienthal fretted, "and would certainly cast serious doubt upon the impartiality and value of

the public utility digest upon which I have spent several years of very hard work, and in which the publishers have an investment of nearly $400,000." See David E. Lilienthal to Charles W. Hadley, July 24, 1931, 2, box 2, subseries 2, folder 2-19, Harry R. Booth Papers.

10. A copy of the withdrawal order (dated July 28, 1931) can be found in box 2, subseries 2, folder 2-14, Harry R. Booth Papers. For Ickes's own battles with Insull, see Linda J. Lear, *Harold L. Ickes: The Aggressive Progressive, 1874–1933* (New York: Garland, 1981), 332–42.

11. Swidler's employment in the U.S. Department of the Interior began on May 31, 1933.

CHAPTER 3. WORKING IN WASHINGTON, D.C.

1. The Brookings Institution, founded in 1916, is a nonprofit organization devoted to the nonpartisan study of public policy.

2. Abe Fortas (1910–1982) was the assistant chief of the legal division of the Agricultural Adjustment Administration (1933–34), assistant director of the Securities and Exchange Administration's public works division (1938–39), general counsel of the Public Works Administration (1939–40), director of the U.S. Interior Department's division of power (1941–42), undersecretary of the U.S. Interior Department (1942–46), and associate justice of the Supreme Court (1965–69).

3. In 1933, the National Industrial Recovery Act (NIRA) created the National Recovery Administration (NRA) to develop and administer a code system that would regulate the production, prices, and labor relations within specific industries. The NRA was declared unconstitutional by the Supreme Court in *Schechter Poultry Corporation v. United States*, 295 U.S. 495 (1935). As part of the NRA, the Public Works Administration (PWA) was allocated $3.3 billion to stimulate the economy by boosting industrial production and by increasing employment through public works projects. It was incorporated into the Federal Works Agency in 1939, which was abolished in 1949. The NIRA is located at 48 Stat. 195.

4. Nathan R. Margold (1899–1947) served as an assistant U.S. attorney (1925–28), special counsel for the New York Transit Commission (1928–29), special counsel for the National Association for the Advancement of Colored People (1930–33), and solicitor of the U.S. Interior Department (1933–42). Charles Fahy (1892–1982) served as an assistant solicitor in the U.S. Interior Department (1933–35), general counsel of the National Labor Relations Board (1935–40), solicitor general of the United States (1941–45), and a judge on the Court of Appeals for the District of Columbia (1949–79).

5. Louis R. Glavis (1883–1971) worked in the U.S. Interior Department from 1904 to 1909, and from 1933 to 1936; during the latter period he was a special investigator for Ickes.

6. The "diary" Swidler is referring to is *The Secret Diary of Harold L. Ickes: The First Thousand Days*, by Harold Ickes (New York: Simon and Schuster, 1953). J. Howard Marshall (1902–1995) was a member of the Petroleum Administrative Board and assis-

tant solicitor in the U.S. Interior Department (1933–35). He also served as the chief counsel of the Petroleum Administration for War (1941–45). The article Swidler mentions by Marshall and Meyers is "Legal Planning of Petroleum Production," *Yale Law Journal* 41 (1931–32): 33–68.

7. Lilienthal's nomination to the TVA encountered some rough political sailing, but was confirmed by the Senate on June 10, 1933. See *Congressional Record—Senate,* 73rd Cong., 1st sess. (June 3, 1933), 4931; ibid. (June 10, 1933), 5624. Before serving as TVA chairman from 1933 to 1938, Arthur E. Morgan (1878–1975) was a nationally respected hydraulic engineer and president of Antioch College (1921–33). See the fine biography by Roy Talbert Jr., *FDR's Utopian: Arthur Morgan of the TVA* (Jackson: Univ. Press of Mississippi, 1987). Harcourt A. Morgan (1867–1950) was an expert in agricultural management and president of the University of Tennessee (1919–33) prior to his tenure as a TVA director (1933–48).

8. William A. Sutherland (1896–1987) was legal secretary to Supreme Court Justice Louis Brandeis (1917–19) and an attorney for the Federal Trade Commission (1919–20). He subsequently became TVA general solicitor (1933–34).

9. The Cosmos Club is a Washington, D.C., social club founded in 1878.

10. Wilson Dam, located in northern Alabama and constructed during World War I, was originally designed to provide electricity for two government-operated nitrate plants. The plants and dam were often called the "Muscle Shoals" project because they were located near the Muscle Shoals section of the Tennessee River. The Wilson Dam complex was turned over to the TVA on Sept. 1, 1933.

11. On Nov. 7, 1933, Ickes wrote to Swidler: "I have your letter of November 7. I have not seen you because there was no good purpose to be served and, as you know, my time is very fully occupied. You chose to accept an appointment with the Tennessee Valley Authority without even consulting me but, of course, you were well within your legal rights in doing so. I may say, however, that in my opinion, I was entitled to some consideration in the premises. And while, as I say, you had the right to do what you did, one might raise a question as to the manner of your doing it and as to the propriety of the Tennessee Valley Authority in enticing away a member of my staff without consulting me." This note is located in Swidler's personnel file, a copy of which is in the editor's possession, pursuant to request number 2443, Jan. 2, 2000, under the U.S. Freedom of Information Act. Swidler's file is approximately 290 pages.

Part II. Initial Work for the TVA

1. Electric power production is usually composed of three components: (1) the generation system involves the facilities and equipment that actually make electricity; (2) the transmission system encompasses the high-voltage lines that convey electricity from the point of generation to a local distributor; and (3) the distributor, or distribution system, includes the lines and equipment (for example, transformers) that reduce the voltage of electricity and transmit it to end-users (consumers). The TVA typically concerned itself with generation and transmission, leaving distribution to municipalities.

2. *Ashwander v. TVA,* 297 U.S. 288 (1936); *Tennessee Electric Power Company v. Tennessee Valley Authority,* 21 F.Supp. 947 (1938); *Tennessee Electric Power Company v. TVA,* 306 U.S. 118 (1939).

3. Joseph C. Swidler and Robert H. Marquis, "TVA in Court: A Study of TVA's Constitutional Litigation," *Iowa Law Review* 32 (Jan. 1947): 296–326. For Swidler's assessment of his limited role in preparing the TVA's constitutional litigation, see the interview of Joseph Swidler, Oct. 29, 1969, 23, in the collection entitled "An Oral History of the Tennessee Valley Authority," a series of interviews conducted by the Oral History Research Office of the University of Memphis between 1969 and 1984.

4. "Two Young Men Credited in Settling Power Issue," *NYT,* Feb. 5, 1939, 25.

CHAPTER 4. BRINGING POWER TO THE TENNESSEE VALLEY

1. Llewellyn Evans (1882–1975), an inventor, engineer, astronomer, and artist, was the chief electrical engineer for the Wilson Dam (1933–37) and chief consulting electrical engineer for the TVA until 1952.

2. By 1941, after various revisions, the wholesale rates for TVA power ranged from 3.9 to 5.8 mills per kilowatt-hour. See Pritchett, *The Tennessee Valley Authority,* 84. Edward Falck (1911–1996) served as the director of rates and research for the TVA (1933–37), and was later appointed deputy director (1943–44) and director (1944–45) of the Office of War Utilities. One of the initial catalysts for the 1938 congressional investigation of the TVA was the so-called "Berry Affair." Chairman Morgan believed that George Berry, a Roosevelt supporter and local landowner, was trying to defraud the TVA (he had filed a claim against the TVA, seeking compensation for property damages caused by flooding in 1936). Lilienthal and Harcourt Morgan, though admitting that Berry's motives and arguments were suspect, nevertheless thought that a settlement could be reached. Viewing the matter differently, Chairman Morgan accused Lilienthal and (Harcourt) Morgan of acting unethically (Chairman Morgan implied that the two might be complicitous in the attempt to commit fraud). Chairman Morgan's refusal to provide evidence to support this accusation (and others) prompted Roosevelt to dismiss him from the TVA on Mar. 23, 1938. The hearings, which lasted from May until December of 1938, evolved into a detailed examination of the TVA's operations. See U.S. Congress, *Hearings before the Joint Committee on the Investigation of the Tennessee Valley Authority* (hereafter *TVA Hearings*), 75th Cong., 3rd sess., parts 1–14, 1938.

3. The initial retail rates for TVA power were 3 cents for the first 50 kilowatt-hours, 2 cents for the next 150 hours, 1 cent for the next 200, and 4 mills for the remaining power used during a month, up to 1400 kilowatt-hours; after that point, the rate increased to three-fourths of a cent. Prior to the introduction of TVA power, residents of the Tennessee Valley were consuming an average of 55 kilowatt-hours per month. TVA administrators hoped that lower rates would boost that figure to 100 kilowatt-hours per month. See Pritchett, *The Tennessee Valley Authority,* 84.

4. TVA records indicate that contracts between the TVA and Tupelo, Mississippi, were signed on Oct. 27 and Nov. 13, 1933. The document Swidler worked on for two-and-a-half months—the one that was signed at the Tupelo ceremony in early February of 1934—was a revised wholesale rate schedule that was not in place when the initial contract had been negotiated. See TVA, *Annual Report for the Tennessee Valley Authority for the Fiscal Year Ended June 30, 1934*, 30, 60. The Electric House and Farm Authority (EHFA) increased demand for electricity by persuading appliance manufacturers and dealers to reduce the prices of their products so that rural residents could afford to purchase them (and thereby use more electricity). Roosevelt, at Lilienthal's urging, created the EHFA on Dec. 19, 1933, by Executive Order 6514; it was chartered as a Delaware corporation on Jan. 17, 1934. On Aug. 12, 1935, the EHFA ceased as a Delaware corporation by Executive Order 7139, and Congress re-incorporated it as a subsidiary of the Reconstruction Finance Corporation. It was liquidated by Executive Order 9256 on Oct. 13, 1942.

5. Swidler is quoting from Lilienthal's published journals. See Lilienthal, *Journals,* 1:39. Wendell Willkie (1892–1944), an Indiana-born lawyer and utilities entrepreneur, began as a Wilsonian Democrat, then ran unsuccessfully as the Republican nominee for president in 1940. He was the president of the Commonwealth & Southern Corporation from 1933 to 1940.

6. Lilienthal, *Journals,* 1, quote at 40.

7. Ibid., 712, quote at 713.

8. Ibid., 713, quotes at 714, 715. According to the final agreement with Willkie, municipalities would have to purchase distribution properties from Commonwealth & Southern, except for some cities in Mississippi, which already operated their own distribution systems. See TVA, *Annual Report, 1934,* 24–26.

9. A copy of the Jan. 4 agreement is reprinted in *TVA Hearings,* 166–73.

10. For the influence of state legislation on subsequent TVA litigation, see *TEPCO v. TVA,* 306 U.S. 118, 141 (1939).

11. The TVA produced and sold three types of power. "Firm" power was available to customers upon demand without restriction. "Interruptible" power was unavailable for specified periods within a given calendar year. "Secondary" power was assured for a given proportion (for example, 75 percent) of a specified period (commonly ten years), but there was no guarantee of power for any certain year within that period. See Twentieth Century Fund, *Electric Power and Government Policy: A Survey of the Relations between the Government and the Electric Power Industry* (New York: Twentieth Century Fund, 1948), 619.

12. Swidler is quoting from his own memorandum, as it is quoted in Thomas McCraw, *TVA and the Power Fight, 1933–1939* (Philadelphia: J. P. Lippincott, 1971), 68. Ickes asserted that the PWA's mandate permitted it to provide loans and grants to municipalities to build their own distribution systems. Though private utility companies argued that using PWA funds for such purposes was unconstitutional, the Supreme Court ruled against them in 1938. See *Alabama Power Company v. Ickes,* 302 U.S. 464 (1938).

13. *TVA Hearings,* quote at 168.

CHAPTER 5. THE TVA BECOMES A POWER SYSTEM

1. William I. Nichols (1905–) was the director of electrical development for the TVA (1934–37). During World War II he worked in the Office of War Utilities.

2. Walton Seymour (1909–1979) conducted economic studies and negotiated power contracts for the TVA (1934–47). He was the director of the division of power and program in the U.S. Department of the Interior (1947–50) and a consultant to the Federal Power Commission (1963–66). Edward Muir was affiliated with the TVA from 1934 until 1947, eventually becoming the chief of the power contracts division. Swidler was likely referring to James E. Watson, an evaluation engineer for the TVA during the late 1930s.

3. Brownsville, the one city in western Tennessee that elected not to purchase TVA power in 1935, signed a contract with the TVA on Dec. 31, 1938.

4. Monsanto was incorporated in 1933 and remains one of the largest agrochemical concerns in the world.

5. TVA and Monsanto signed a contract on May 15, 1936. The contract is reprinted in TVA, *Annual Report, 1936,* 195–203, quote at 196.

6. The quoted language was added by Swidler himself to the legislation. See the 1935 amendments to the TVA Act, located at 49 Stat. 1075, quote at 1077.

7. Alcoa was incorporated in 1888 and virtually created the market for aluminum, with annual sales exceeding $10 billion by the 1990s.

8. 48 Stat. 58, quote at 69.

9. The standard work on the Morgan-Lilienthal conflict is Thomas McCraw, *Morgan vs. Lilienthal: The Feud within the TVA* (Chicago: Loyola Univ. Press, 1970).

10. The statutory provision allowing the TVA to exchange power for dam sites had expired by the spring of 1936. Moreover, during that same spring, the U.S. House of Representatives failed to appropriate any funds that would have permitted the TVA to purchase the Fontana site or built a dam. Chairman Morgan wanted the TVA board to send a letter to the U.S. Senate asking for an appropriation to cover, at a minimum, the costs of building the Fontana Dam. It was this letter that the board, during Chairman Morgan's absence on June 2, 1936, voted not to send, which indefinitely postponed the Fontana Dam project. See *TVA Hearings,* 373. John ("Jack") B. Blandford Jr. (1897–1972) was an assistant to David Lilienthal (1933–34) before becoming the coordinator and secretary of the TVA board (1934–37) and TVA general manager (1937–39). He later served as the director of the United Nations Relief and Works Agency for Palestine Refugees (1951–53).

11. *TVA Hearings,* quote at 371.

12. George W. Norris (1861–1944), a Republican, was a U.S. representative (1903–13) and a U.S. senator (1913–43) from Nebraska. He was perhaps best known for championing the legislation that eventually resulted in creation of the TVA (both Norris Dam and the town of Norris, Tennessee, were named after him). The "yardstick" concept referred to the use of public power rates as a basis to assess the equity of private power rates. Such comparisons produced ambiguous data: Larger municipal plants commonly charged lower rates than private ones, while smaller munici-

pal plants usually charged higher rates than their private sector counterparts. See McCraw, *TVA and the Power Fight,* 164.

13. Both Lilienthal and Swidler believed that Chairman Morgan expressed too much sympathy for Wendell Willkie's views. For that reason, they tended to believe that support of Willkie represented support of Chairman Morgan as well.

14. The 1936 contract with Alcoa, negotiated after the Fontana Dam discussions had been halted, was a reciprocal power agreement: Alcoa would supply electricity for the TVA's construction of Hiwassee Dam (initially called Fowler Dam), while the TVA would supply—via a new transmission line—an equal amount of electricity from Norris Dam to Alcoa's operations in Alcoa, Tennessee. This obviated the necessity for the TVA to build a costly, temporary transmission line from one of its power generators to the Hiwassee site. A year later (1937), Alcoa began purchasing additional electricity from the TVA, which was supplied by the Norris Dam transmission line. See TVA, *Annual Report for the Tennessee Valley Authority for the Fiscal Year Ended June 30, 1937* (Washington, D.C., 1937), 150–53. Gabriel O. Wessenauer (1906–1990) worked for the TVA as a hydraulic engineer (1935–37), power supply engineer (1937–39), assistant manager of power (1940–44), and manager of power (1944–70). Roland A. Kampmeier (1910–) served the TVA as an assistant hydraulic engineer (1933–39), director of power utilization (1941–52), and director of power supply (1952–60).

15. The American Electric Power Company was incorporated in 1906 as the American Gas and Electric Company. Philip Sporn (1896–1978) was an Austrian-born, American-educated electrical engineer known for his many innovations concerning the production and distribution of electricity. He was associated with the American Electric Power Company for several decades, serving as its president from 1947 to 1961.

16. James P. Growdon (1884–1969) worked for Alcoa as a hydraulic engineer (1925–54); he is credited with the development of the sloping earth-core rock-fill dam.

17. TVA, *Annual Report, 1937,* 309–12. Victor Chemical Works was incorporated in 1902, and by the 1930s was one of the country's largest producers of phosphates. It became a division of the Stauffer Chemical Company in 1959. The TVA contract with Victor Chemical was signed on July 2, 1937. See *TVA Hearings,* 839.

CHAPTER 6. EXPANSION OF THE TVA POWER MARKET

1. TVA, *Annual Report, 1934,* quote at 22.

2. Ibid., quotes at 23.

3. The Tri-Cities area is often subsumed under the "Quad-Cities" rubric that designates Florence, Muscle Shoals, Sheffield, and Tuscumbia, Alabama.

4. On Mar. 4, 1938, Lilienthal wrote to Willkie, as well as the presidents of the Mississippi Power Company, the Alabama Power Company, and TEPCO, offering to purchase specified properties on the basis of their "'actual legitimate cost . . . with due consideration for loss in value resulting from their past use.'" Willkie responded that the sale of the properties would have to be calculated according to their value as "'going concerns.'" Both individuals are quoted in "Memorandum of the Positions of

the Parties in the Recent Negotiations between TVA and the Commonwealth and Southern Corporation for the Purchase of Certain Utilities Properties," located in the collection labeled "Board of Directors: David E. Lilienthal Files," Forrest Allen folder, National Archives and Records Administration, Southeast Region, East Point, Georgia.

5. 48 Stat. 58, quote at 64. While there was never formal concurrence over the TVA's territorial boundaries, the TVA eventually agreed to restrict its expansion voluntarily.

6. Julius "Cap" Krug (1907–1970), an expert in the field of public utilities, eventually became U.S. secretary of the interior (1946–49). He also served—as Swidler did—in the War Production Board (1942–44). Disagreement over the value of TEPCO properties became clear at a meeting between Lilienthal and Willkie on June 13, 1938. Negotiations came to a virtual standstill until Swidler and Krug replaced Lilienthal later that year. Lilienthal contracted undulant fever on Aug. 31, 1938. See Lilienthal, *Journals,* 1:118.

7. Other sources suggest that some of the negotiations among Willkie, Krug, and Swidler were less than cordial. See the interview of States Rights Gist Finley (July 23, 1971, 7) in "An Oral History of the Tennessee Valley Authority." The first article Willkie published in the *Saturday Evening Post* was "Idle Money, Idle Men," which appeared in the June 17, 1939, issue. Since most of Swidler and Krug's negotiations with Willkie took place in December of 1938 and January of 1939, it seems likely that Swidler saw this article during a later visit to Willkie's office. It was reprinted in Wendell Willkie, *This Is Wendell Willkie: A Collection of Speeches and Writings on Present-Day Issues* (New York: Dodd, Mead, 1940), 83–100. An article *about* Willkie did appear in the winter of 1939. See Alva Johnston, "The Man Who Talked Back," *Saturday Evening Post,* Feb. 25, 1939, 210ff.

8. The TVA and Willkie reached a tentative agreement on Feb. 4, 1939, which the TVA board approved in principle the following day. The general outlines of that understanding were initially communicated to Lilienthal during a telephone call from Willkie (Lilienthal was still recovering from undulant fever). See telephone transcript between David Lilienthal and Wendell Willkie, Feb. 4, 1939, David E. Lilienthal Papers, box 92; Lilienthal, *Journals,* 1:85. Ultimately, the federal government (via the TVA) paid Willkie $44,728,300, while the balance (almost $34,000,000) was paid by local distributors. See "TVA Takes Title to Power for State of Tennessee," *NYT,* Aug. 16, 1939, 1.

9. Bernard F. Weadock (1884–1947) was a law partner of Wendell Willkie (1929–32) and vice-president and managing director of the Edison Electric Institute (1933–39).

10. John Lord O'Brian (1874–1973) was a Republican in the New York Assembly (1907–09), U.S. attorney (1909–14), and assistant to the U.S. attorney general (1929–33). He was also a special counsel in the U.S. Justice Department during the 1930s; it was in that position that he argued *Ashwander* before the Supreme Court. He was subsequently general counsel for the War Production Board (1941–44).

11. The precise date for the meeting that O'Brian, Swidler, Krug, Weadock, and Willkie attended cannot be determined. Swidler's travel records indicate that he was in Washington, D.C., on Mar. 13–14, 1939, to discuss "proposed legislation." The

meeting to which Swidler refers might well have occurred during that visit. For Swidler's travel records, see unorganized 1999 addition to the Joseph Swidler Papers (hereafter SP), Library of Congress, accession 22,094.

12. The negotiations to which Swidler is referring took place in April of 1939. The date on the official contract is May 12, 1939. See "Willkie and TVA Draft Contract," *NYT,* Apr. 13, 1939, 33; TVA, *Annual Report of the Tennessee Valley Authority for the Fiscal Year Ended June 30, 1940* (Washington, D.C., 1940), 87. L. J. Wilhoite (1887–1960) was chairman of the Chattanooga Electric Power Board (1938–60). States Rights Finley (1898–1980) was the general superintendent of the Chattanooga Electric Power Board (1938–63) and president of the Tennessee Valley Public Power Association (1947–48). Some of the fiercest debates over private versus public power occurred in Chattanooga.

13. Lilienthal, *Journals,* 1:120. Recent biographies of Lilienthal and George Norris by Steven Neuse and Richard Lowitt, respectively, as well as Lilienthal's own journals, suggest that Lilienthal was less enthusiastic about lobbying Congress than Swidler's account implies. See Neuse, *David E. Lilienthal,* 118; Richard Lowitt, *George W. Norris: The Triumph of a Progressive, 1933–1944* (Urbana: Univ. of Illinois Press, 1978), 278–82; Lilienthal, *Journals,* 1:119–21.

Part III. Washington, D.C., and Wartime Service

CHAPTER 7. THE DEPARTMENT OF JUSTICE AND THE WAR PRODUCTION BOARD

1. Theodore Parker (1889–1944) was the TVA's chief construction engineer from 1935 to 1943. According to Swidler's travel records, the TVA trip lasted from Apr. 9 to Apr. 28, 1940. See unorganized 1999 addition to SP.

2. Maury Maverick (1895–1954), a Democrat, was a U.S. representative from Texas (1935–39), mayor of San Antonio (1939–41), and director of the government division (1941–43) and vice-chairman (1943–46) of the War Production Board.

3. Swidler's travel records indicate that the group stopped in Nashville, Memphis, Austin, El Paso, Phoenix, and Boulder City (Nevada) prior to visiting Los Angeles. After Los Angeles, they continued on to Visalia (California), Fresno, San Francisco, Redding, Portland, Spokane, Williams (California), and then back to Los Angeles. On their return to Knoxville, they stopped in Sweetwater (Texas), Abilene, and Memphis. See unorganized 1999 addition to SP.

4. The Federal Power Commission (FPC) was created as a cabinet committee on June 10, 1920, as part of the Federal Water Power Act (41 Stat. 1063). The Federal Power Act of 1930 (46 Stat. 797) granted the FPC independent status, and its regulatory powers were broadened under Titles II and III of the Public Utility Holding Company Act of 1935 (49 Stat. 803) to include interstate sales of electricity and gas (Title II is known separately as the Federal Power Act of 1935). On Aug. 4, 1977, the FPC was terminated by the U.S. Department of Energy Organization Act (91 Stat. 578); many of its functions were transferred to the Federal Energy Regulatory Commission (FERC) within the U.S. Department of Energy. The only remark from Lilienthal's *Journals* that supports Swidler's statement about Lilienthal's opinion of

him is the following: "The boys in the TVA power staff, plus (General Counsel) Swidler, are just reluctant to do things differently than they have been doing them, i.e., financing capital expansion by annual appropriations. They believe that somehow things will work out, after the usual annual fight. I wonder if they have been operating men too long to have the drive and creativeness that meeting a new situation requires. I like them, and respect their great devotion (to the death) to TVA. But what is needed now is a realistic recognition of a change of circumstances." See *The Journals of David E. Lilienthal: Venturesome Years, 1950–1955,* vol. 3 (New York: Harper, 1966), 606.

5. Swidler was promoted from assistant general counsel (a title he had held since Feb. 1, 1937) to solicitor on Aug. 9, 1939. James Lawrence Fly (1898–1966) served as special assistant to the U.S. attorney general for antitrust cases (1929–34), general solicitor and general counsel for the TVA (1934–39), and chairman of the Federal Communications Commission (1939–44). William C. Fitts Jr. (1905–1954) served the TVA as assistant general solicitor (1935–36), assistant general counsel (1936), solicitor (1937–39), and general counsel (1939–44). Charles J. McCarthy (1907–1995) joined the TVA in 1934 and succeeded Swidler as general counsel (1957–67).

6. Francis "Frank" Shea (1915–1978) served in the legal opinion section of the Agricultural Adjustment Administration (1933–35) and as general counsel of the Puerto Rico Reconstruction Administration (1935–36). Shea was also assistant U.S. attorney general (1939–44) and an associate prosecuting attorney at the Nuremberg war trials in 1945. Within the U.S. Department of Justice, the Alien Property Bureau (1934–41) was the predecessor to the Alien Property Division (1941–42), whose functions were taken over by the Office of Alien Property Custodian (1942–46).

7. The contract authorizing the loan of Swidler to the U.S. Department of Justice is dated Aug. 1, 1941. The Foreign Funds Control Office was established in the Office of the U.S. Secretary of the Treasury in April of 1941. Henry Morgenthau Jr. (1891–1967) was chairman of the Federal Farm Board (1933), chairman of the Farm Credit Administration (1933–34), undersecretary (1934) and U.S. secretary of the treasury (1934–45).

8. Dean Acheson (1893–1971) was appointed undersecretary of the U.S. treasury in 1933, and later served as U.S. secretary of state (1949–53). He played a major role in shaping the American policy of "containing" communism in the post–World War II period.

9. Leo T. Crowley (1890–1972) was chairman of the board of directors of the Federal Deposit Insurance Corporation (1934–45), director of the Office of Alien Property Custodian (1941–43), head of the Office of Economic Warfare (1943), and director of the Foreign Economic Administration (1943–45).

10. The War Production Board (WPB) was created by Executive Order 9024 on Jan. 16, 1942. Its primary function was to coordinate the production and distribution of resources during World War II. By early 1943, the Office of War Utilities (located in the WPB's power division) had jurisdiction over any matter relating to: (1) producing, transmitting, and distributing electricity, manufactured gas, water, and central steam heating; (2) transmitting and distributing natural gas; (3) furnishing of

communications except radio; and (4) redistributing materials and equipment owned by, in the possession of, or held for the account of, persons engaged in such activities and services. See *Federal Register* 8 (Feb. 16, 1943): 2007. Swidler returned to the TVA in February of 1942. His loan to the WPB officially began on Aug. 4, 1942. Herbert Marks (1907–1960) was an assistant general counsel of the TVA (1934–39) and the first general counsel of the Atomic Energy Commission (1947).

11. Swidler was stationed in Kansas City from December of 1942 through January of 1943. His loan to the WPB expired on Jan. 24, 1943.

12. The phrase "Manhattan Project" refers to the American effort to construct an atomic bomb during World War II. For that purpose, a site was built in Oak Ridge, Tennessee, to separate uranium-235 from the more common uranium-238.

CHAPTER 8. THE SEABEES AND THE CONTRACT TERMINATION OFFICE

1. Rear Admiral Ben Moreell sought permission to form a construction force (the Seabees) on Dec. 28, 1941; he received permission to do so on Jan. 5, 1942. The Seabee division was eventually comprised of 8,000 officers and 250,000 recruits. Its name was derived from the initials "CB" (construction battalions). Swidler was a Seabee from June 10, 1943 (the date of his initial enlistment) until Nov. 15, 1943.

2. Camp Peary, the primary base for training Seabees, was a ten thousand–acre facility in Williamsburg, Virginia. It became operational in November of 1942, and, by the end of the war, it had trained approximately 175,000 recruits. The camp was turned over to the CIA in 1951.

3. The SSCRA legislation is located at 54 Stat. 1178 and 56 Stat. 769.

4. Swidler was stationed in the Contract Termination Office from Nov. 15, 1943 until Oct. 1, 1945. Lewis Strauss (1896–1974), a professional banker, was a member (1946–50) and chairman (1953–58) of the Atomic Energy Commission (AEC). Strauss held a reserve commission in the navy, and was given the honorific title of rear admiral at the end of his wartime service.

5. Swidler is likely referring to John M. Young (1901–1978), one of the organizers and first directors of Morgan Stanley and Company. Young worked in the Navy Department's Office of Procurement and Material (1943–45).

6. James Forrestal (1892–1949) was an investment banker appointed as an administrative assistant to Franklin Roosevelt in 1940. He served as undersecretary (1940–44) and secretary of the navy (1944–47). He also served as the first U.S. secretary of defense (1947–49) after the War Department and Navy Department were joined in 1947.

7. J. Robert Oppenheimer (1904–1967) was a theoretical physicist who, as the director of the Los Alamos Weapons Laboratory (1942–45), helped build the first atomic bomb. After the war, Oppenheimer served as the director of the Institute for Advanced Study at Princeton University (1947–66) and chairman of the Atomic Energy Commission's General Advisory Committee (1947–52). The Atomic Energy Commission (AEC) was created by the Atomic Energy Act of 1946 (60 Stat. 755) for the purpose of licensing and regulating the civilian uses of nuclear power. Its functions were transferred to the Nuclear Regulatory Commission under the provisions of

the Energy Reorganization Act of 1974 (88 Stat. 1233). Lewis Strauss, as chairman of the AEC, supported an ultimately successful effort to strip Oppenheimer of his security clearance in 1953. At issue was Oppenheimer's association with communists and his belief that scientific information and materials (including radioactive isotopes) should be shared with America's allies. Disagreement over this latter point initially fueled Strauss's seemingly personal vendetta against Oppenheimer. See Philip M. Stern, *The Oppenheimer Case: Security on Trial* (New York: Harper, 1969), 37, 114, 425.

8. David E. Lilienthal, *The Journals of David E. Lilienthal: The Atomic Energy Years, 1945–1950,* vol. 2 (New York: Harper, 1964), 89.

9. Swidler worked on the report from Oct. 1–Dec. 6, 1945, after which he was discharged from the armed services.

10. Swidler returned to the TVA on Dec. 7, 1945.

11. Swidler's divorce was finalized on July 28, 1943. He married Gertrude Tyrna on May 5, 1944. In addition to a daughter, Swidler and Gretchen also had a son, Mark, who was born in 1948.

Part IV. General Counsel for the TVA

1. "Eisenhower Points to the TVA as 'Creeping Socialism' Example," *NYT,* June 18, 1953, 1; Dwight D. Eisenhower, "President's News Conference of June 17, 1953," in *Public Papers of the Presidents of the United States: Dwight D. Eisenhower, 1953–1960,* vol. 1 (Washington, D.C., 1954), 433; Aaron Waldavsky, *Dixon-Yates: A Study in Power Politics* (New Haven: Yale Univ. Press, 1962), 139.

2. The amendments that made the TVA self-financing were approved on Aug. 6, 1959. See 73 Stat. 280.

3. For the view that Lilienthal remained ideologically consistent, see Neuse, *David E. Lilienthal,* 256.

CHAPTER 9. SERVICE AS TVA GENERAL COUNSEL

1. Marguerite Owen (1896–1983), the author of a 1929 pamphlet for the National League of Women Voters entitled "Muscle Shoals and the Public Welfare," was appointed to the TVA in 1933. She served in the general manager's office (Washington representative) from 1934 until her retirement in 1966.

2. Marguerite Owen, *The Tennessee Valley Authority* (New York: Praeger, 1973). Gordon R. Clapp (1905–1963) began his TVA service in 1933 and was chairman from 1947 to 1954. He was later president of the Development and Resources Corporation (1955–63).

3. Bronson M. Cutting (1888–1935), a Republican, was a U.S. senator from New Mexico (1927–35). J. Lister Hill (1894–1984), a Democrat, was a U.S. representative (1923–38) and a U.S. senator (1938–69) from Alabama. Hugo L. Black (1886–1971), a Democrat, was a U.S. senator from Alabama (1926–37) and an associate justice of the Supreme Court (1937–71). Kenneth D. McKellar (1869–1957), a Democrat, was a U.S. representative (1911–17) and a U.S. senator (1917–53) from Tennessee. John E. Rankin (1882–1960), a Democrat, was a U.S. representative from Mississippi (1921–53); he was also a coauthor of the bill creating the TVA. C. Estes Kefauver

(1903–1963), a Democrat, was a U.S. representative (1939–49) and a U.S. senator (1949–63) from Tennessee. Albert A. Gore Sr. (1907–1998) was a U.S. representative (1939–53) and a U.S. senator (1953–71) from Tennessee.

4. David E. Lilienthal, *TVA: Democracy on the March*, rev. ed. (New York: Harper, 1953), quote at xxiii.

5. According to the TVA's enabling legislation, the corporation's "principal office" was supposed to be located in the "immediate vicinity" of Muscle Shoals. That, however, was never the case; the TVA's administrative offices have always been located in Knoxville, Tennessee. Though Tennessee's congressional delegation was unsuccessful in its attempts to amend the enabling legislation so that it would officially recognize Knoxville as the TVA's headquarters, the federal courts ruled in 1981 that the headquarters could remain there. Previous attempts supported by TVA officials to move the headquarters to Muscle Shoals were likely based on the belief that Alabama's congressional delegation would be more supportive of the TVA's interests than the Tennessee delegation. See 48 Stat. 58, quotes at 64; *Alabama ex rel. Graddick v. Tennessee Valley Authority*, 636 F.2d 1061 (1981).

6. At least one scholar has addressed the issue of the TVA's increasing reliance on steam generators during the late 1940s and 1950s. See Erwin C. Hargrove, *Prisoners of Myth: The Leadership of the Tennessee Valley Authority, 1933–1990* (Princeton: Princeton Univ. Press, 1994), especially chap. 6.

7. Herbert Vogel (1900–1984), a civil engineer and a graduate of the United States Military Academy, was appointed by Eisenhower as chairman of the TVA in 1954, a post he held until 1964. Harry Curtis (1884–1963) was a chemical engineer who served on President Coolidge's Muscle Shoals Commission; he later became dean of the University of Missouri's College of Engineering (1939–48) and a TVA director (1949–57). Raymond Paty (1896–1957), a former teacher and university administrator, was a TVA director (1952–57). Early opposition to the TVA's construction of steam generator plants was manifested in 1948 by congressional refusal to fund one such plant in Johnsonville, Tennessee. See "House Rejects a TVA Power Plan Requiring Steam-Generator Plant," *NYT*, May 12, 1948, 1.

8. The case Swidler argued before the Supreme Court was *United States ex rel. Tennessee Valley Authority v. Welch*, 327 U.S. 546 (1946). The record indicates that Welch's counsel was McKinley Edwards; all other respondents were represented by George H. Ward.

9. Both Truman and Eisenhower came under public pressure to tighten security regulations concerning government employees. To that end, Truman issued Executive Order 9835 in 1947. It established a federal Loyalty Review Board (under the aegis of the Civil Service Commission) and separate loyalty boards within various government agencies. The Loyalty Review Board, with the assistance of the FBI, investigated the backgrounds of current and prospective federal employees. During the board's first four years, over 3 million individuals were investigated; only 212 were dismissed as security risks. Bowing to the era's anti-communist blustering, Truman, then Eisenhower, issued additional executive orders (EO 10241 in 1951 and EO 10450 in 1953) that relaxed the evidentiary standards used by officials to bring

charges of disloyalty. Past, not just present, activities could now be adduced to illustrate an individual's alleged disloyalty. Some of these new criteria were used to justify the charges brought against Swidler. For data concerning loyalty investigations, see Thomas S. Arms, *Encyclopedia of the Cold War* (New York: Facts on File, 1994), 136–38, 239–40.

10. The specific charges, and Swidler's written rebuttal of them, can be found in Swidler to Harry L. Case, Sept. 20, 1951, unorganized 1999 addition to SP. Case was the TVA's director of personnel (1948–55) and acting chair of its loyalty board during Swidler's investigation. For information on the formation of the TVA's loyalty board, see "TVA Loyalty Board to Aid Menhinick," *Knoxville News-Sentinel*, Dec. 12, 1947, located in the TVA Corporate Library, Knoxville, Tennessee.

11. Swidler's account omits a few details. In 1947, during hearings over the nomination of Gordon Clapp as TVA chairman, Senator Kenneth McKellar intimated that Swidler and two other TVA officials (Arthur Jandrey and E. B. Schultz) were communists. Rather than produce any evidence to substantiate these charges, McKellar apparently hoped that his accusations would undermine Clapp's chances of becoming TVA chairman. McKellar's animus was actually directed less at Clapp or Swidler, than at David Lilienthal, who supported Clapp's nomination. McKellar was incensed by Lilienthal's longstanding refusal to grant him patronage powers over TVA staffing. He was also angered by Lilienthal's opposition to his choice for the Douglas Dam site. In what seems to have been a retaliatory strike, McKellar raised the specter of communism during Lilienthal's nomination hearings as chairman of the AEC. McKellar failed to derail either Lilienthal's or Clapp's nomination (their hearings overlapped), but his remarks and insinuations might well have provoked enough suspicion to fuel the FBI's subsequent investigation of Swidler. For McKellar's initial comments, see "Communists in TVA, McKellar Believes," *NYT*, Jan. 22, 1947, 1.

12. *The New Republic* began circulation in 1914 and, for many decades, was known for its liberal political and social views.

13. Swidler asserted that the FBI's two primary informants against him were Roxanne J. Paris Hubbert (whom everyone referred to as "Mrs. Paris") and L. B. Bolt Jr. Paris, a resident of Knoxville, Tennessee, had a history of making dubious accusations. Bolt had been a TVA employee until 1946, when the need to employ returning servicemen, including Swidler, required his termination. Swidler documented Bolt's subsequent threat that he would "get" Swidler for dismissing him. See Swidler to Case, Sept. 20, 1951, 2, 28.

14. This particular charge was related to the accusation that Swidler supported positions beneficial to the Soviet Union. In Swidler's written rebuttal, he speculated that Arthur L. Davis was probably the patent attorney who served as the FBI's informant. Davis was employed by the TVA from Feb. 13, 1934, to Apr. 30, 1948. The former U.S. commissioner of patents with whom Swidler consulted was Casper W. Ooms. See Swidler to Case, Sept. 20, 1951, 24, 25.

15. The so-called Soviet purges, launched by Joseph Stalin between 1936 and 1938, initially targeted senior members of the Communist Party, but then expanded to

include army officers, union officials, intellectuals, and ordinary citizens. Documented by newsreels, the judicial proceedings against these individuals were largely unsuccessful attempts to disguise Stalin's exercise of totalitarian power. The "show trials" alienated liberals and the non-communist left from the Soviet Union, and were used by political reactionaries in the United States and elsewhere to discredit anyone or group that supported communism.

16. The Spanish Civil War (1936–39) pitted the "Nationalists" against the "Loyalists." The latter supported the democratically elected government of the Spanish Republic, a coalition that included socialists and communists. Since the Soviet Union endorsed the Loyalist cause, Americans who did likewise were often charged with being communists or communist sympathizers.

17. The National Lawyers Guild, founded in 1937, is dedicated to the defense of civil liberties and human rights. It was one of the only organizations that defended individuals accused of disloyalty during the 1950s. For this and other reasons, detractors claimed that the Guild's guiding influence was communism. Swidler was a member of the National Lawyers Guild from 1937 to 1940. The meeting at which he spoke was held on May 31, 1940, in New York City. See Swidler to Case, Sept. 20, 1951, 20.

18. Henry H. ("Joe") Fowler (1908–2000) was a counsel (1934–38) and assistant general counsel (1939) for the TVA prior to his service in the War Production Board (1942–44). Robert Marquis (1913–) was a special attorney for the U.S. Justice Department (1938–39) before becoming assistant general counsel (1947–58), solicitor (1958–67), and general counsel (1967–75) for the TVA.

19. J. Edgar Hoover (1895–1972) served as the director of the FBI for several decades (1924–72). In his quest to ferret out political subversives, he often ignored or violated civil liberties. Joseph R. McCarthy (1908–1957), a Republican, was a U.S. senator from Wisconsin (1947–57). McCarthy gained national prominence in 1950 when he claimed that the U.S. State Department employed known communists. For the next several years, he used the powers of his office and the Senate's Government Operations Committee to identify, question, and harass anyone or group whom he suspected of supporting communism. His tactics gave rise to the term "McCarthyism" and helped initiate a period of widespread suppression of civil liberties.

CHAPTER 10. ENVIRONMENTAL PROTECTION AND A PRIVATIZATION EFFORT

1. Lilienthal, *Journals*, 3:194.

2. Lazard Freres and Company, founded in 1848, is a privately controlled, international investment banking firm. André Meyer (1898–1979) was an investment banker who was the senior partner of Lazard Freres from 1944 until 1977.

3. James P. Pope (1884–1966), a Democrat, was a U.S. senator from Idaho (1933–39) and a TVA director (1939–51).

4. Joseph Volpe Jr. (1913–) was counsel to the Manhattan Project (1943–46), as well as associate counsel (1946–48) and general counsel (1948–51) of the AEC. He also represented the State of Tennessee in its opposition to the Dixon-Yates plan.

CHAPTER 11. THE DIXON-YATES SCHEME

1. First Boston was created in 1934 and became associated with Credit Suisse in 1978. It specializes in international banking and finance.

2. Edgar H. Dixon (1904–1962) was president of the Electric Power and Light Corporation (1944–49) and Middle South Utilities (1949–62). Eugene Yates (1880–1957) was chairman of the Southern Company and vice-president and director of the Alabama Power Company, the Georgia Power Company, and the Gulf Power Company.

3. Contemporaries referred to this arrangement as "replacement power," since the TVA would use the output it received from the MRGC (approximately 500,000–600,000 kilowatts) to "replace" the power it was providing to the AEC. The estimated construction cost of the West Memphis (Arkansas) power plant was $107 million; the First Boston Corporation would have played a key role in financing it.

4. Swidler's last point is one that other opponents of Dixon-Yates also made. It was a fiction, they argued, to think that the TVA could send the power it received from the MRGC as a discrete, identifiable unit to the AEC plants in Oak Ridge and Paducah, then use the kilowatts it "replaced"—that is, use only TVA-generated kilowatts—to augment its Memphis load (as if kilowatts could be labeled "AEC" or "TVA"). For this reason, and in contrast to Swidler's own comments, it was not entirely accurate to assert that, under Dixon-Yates, the TVA would use MRGC power to "lose" or "replace" its AEC load. In reality, the TVA would have to utilize the privately generated power it acquired from the MRGC (as per the MRGC's contract with the AEC) to meet the increased demand of the Memphis area. This was the incursion of private power into the TVA's market that public power advocates opposed and the Eisenhower administration supported.

5. Henry DeWolf Smyth (1898–1986), a Princeton University physicist, was a scientific adviser to the Manhattan Project (1943–45) and a member of the AEC (1949–54). Thomas E. Murray (1891–1961), an engineer, inventor, and successful businessman, was a member of the AEC (1950–57). Eugene M. Zuckert (1911–2000) was an assistant counsel for the Securities and Exchange Commission (1937–40), assistant secretary of the air force (1947–52), and a member of the AEC (1952–54).

6. The Tennessee Valley Public Power Association, which has been active since the mid-1940s, is a nonprofit organization that coordinates benefits, training, and communication among the 110 municipal utilities and the 50 rural electric cooperatives in the TVA service area.

CHAPTER 12. FIRED BY GENERAL VOGEL

1. Arnold R. Jones (1904–1975) was appointed to the TVA board in 1957, a position he held until 1966.

2. Hugh Scott Jr. (1900–1994) was a U.S. representative (1941–45, 1947–59) and a U.S. senator (1959–77) from Pennsylvania. Andrew Schoeppel (1894–1962) was governor of Kansas (1943–47) and a U.S. senator from Kansas (1949–62).

3. United States Senate, *Hearings before the Committee on Interstate and Foreign Commerce (Nomination of Joseph C. Swidler to be a Member of the Federal Power*

Commission), 87th Cong., 1st sess., Apr. 11, May 2, and May 9, 1961, quotes at 9–11 (hereafter *Swidler Hearings*).

4. James Strom Thurmond (1902–) was governor of South Carolina (1947–51) and U.S. senator from South Carolina (1954–). He changed his political affiliation from Democrat to Republican in 1964. Thurmond's questioning of Swidler can be found in *Swidler Hearings*, 24–33.

5. Felix Frankfurter (1882–1965) was an Austrian-born legal scholar who taught at Harvard Law School (1914–39) and served as an associate justice of the Supreme Court (1939–62). Frankfurter used Swidler's departure from the TVA as the focus of a rambling address he gave to the Federal Bar Association in September of 1957. Frankfurter warned that there were "clear questions of policy that are not the concern of the General Counsel." On the other hand, he also cautioned that "if we are going to have the lawyers change with the policymakers we are not going to have law." See untitled speech, Thirty-Seventh Annual Dinner of the Federal Bar Association, Sept. 21, 1957, quotes at 39, 40, reel 128, Felix Frankfurter Papers, Library of Congress. See also, "TVA Aide Quits in Board Change," *NYT*, Sept. 15, 1957, 62.

CHAPTER 13. A TVA BALANCE SHEET

1. Gordon R. Clapp, *The TVA: An Approach to the Development of a Region* (Chicago: Univ. of Chicago Press, 1955).

2. Benjamin V. Cohen (1894–1983), a member of Roosevelt's "Brain Trust," helped draft several pieces of New Deal legislation, including the Securities Exchange Act of 1934 (48 Stat. 881), the Public Utility Holding Company Act of 1935 (49 Stat. 803), and the Fair Labor Standards Act of 1938 (52 Stat. 1060). Among his many positions in government, he served as associate general counsel of the PWA (1933–34), general counsel of the National Power Policy Committee (1934–41), and general counsel of the Office of War Mobilization (1943–45). See n. 3 below for an explanation of Swidler's reference to the "Nine TVAs bill."

3. Though the Norris-Rankin bill was called the "Seven TVAs" bill (see, for example, *U.S. News and World Report*, June 6, 1937, 2), it embraced an eighth agency already in existence, the Mississippi River Commission, which was created on June 28, 1879. Swidler probably referred to it as the "Nine TVAs" bill because of his and Herman's initial assignment to include provisions regulating the operations of the Bonneville Dam. Construction of the Bonneville Dam lasted from 1933 until 1937, but the Bonneville Power Administration, contrary to Swidler's assertion, was not yet a reality in the spring of 1937 when he and Herman worked on what became known as "Norris-Rankin." Moreover, the "interim bill" Swidler mentions was likely part of the Bonneville Power Administration's enabling legislation, which was passed on Aug. 20, 1937 (50 Stat. 731). For analyses of these events, see Lowitt, *George W. Norris*, 200–203, with the "seven TVAs" listed at 202; Funigiello, *Toward a National Power Policy*, 174–93; Lilienthal, *Journals*, 1:278–79; Craufurd D. Goodwin, "The Valley Authority Idea: The Fading of a National Vision," in Erwin C. Hargrove and Paul K. Conkin, eds., *TVA: Fifty Years of Grass-Roots Bureaucracy* (Chicago: Univ. of Illinois Press, 1983), 263–96; and William E. Leuchtenburg, *The FDR Years: On Roosevelt*

and His Legacy (New York: Columbia Univ. Press, 1995), 159–95. Clyde T. Ellis (1908–1980), a Democrat, served in the Arkansas House of Representatives (1933–35), the Arkansas Senate (1935–39), the U.S. House of Representatives (1939–43), and as general manager of NRECA (1943–67). NRECA was founded in 1942 to represent the interests and views of cooperative electric utilities.

4. The TVA's enabling legislation permitted it to "manufacture and sell fixed nitrogen, fertilizer, and fertilizer ingredients at Muscle Shoals." If the TVA did in fact produce these products, it was required to test them under "large-scale practical use." For this purpose, the TVA was authorized to donate fertilizer to agricultural colleges and demonstration farms. Harcourt Morgan, in an apparent deviation from congressional intent, influenced the TVA in becoming a producer of phosphatic, not nitrogenous, fertilizers. See 48 Stat. 58, quotations at 61.

5. The "portrait" Swidler refers to is the 1948 photograph taken by Arnold A. Newman, which was donated to the National Portrait Gallery in 1988 by the Lilienthal family.

6. Swidler's statement about Lilienthal's journals being written with "eventual publication in mind" is only partially true. The early journals were not written with this objective; the later ones probably were. See Neuse, *David E. Lilienthal,* xv–xvii.

7. David E. Lilienthal, *The Journals of David E. Lilienthal: Unfinished Business, 1968–1981,* vol. 7 (New York: Harper, 1983), 253.

8. Henry A. Wallace (1888–1965), a Democrat, served as U.S. secretary of agriculture (1933–40), vice-president (1941–45), and U.S. secretary of commerce (1945–46). He was also the presidential nominee of the Progressive Party in 1948. Charles Beard (1874–1948) was a noted historian and author whose best-known work was the controversial *An Economic Interpretation of the Constitution* (1913). Stuart Chase (1888–1985) was an economist and prolific author who coined the phrase "a New Deal."

9. Lilienthal, *TVA,* xxii.

10. Ibid., 138.

11. Ibid., 151.

12. Ibid., 151, 155.

13. Lilienthal, *Journals,* 2:150. Bernard Baruch (1870–1965) was a financier and statesman who advised every president from Woodrow Wilson to Dwight D. Eisenhower.

14. The Soviet Union detonated its first atomic bomb on Aug. 29, 1949. In addition to Lilienthal, the "special committee" of the National Security Council was composed of U.S. Secretary of State Dean Acheson and U.S. Secretary of Defense Louis Johnson. Though Lilienthal did not agree with all of the committee's recommendations, he signed the final draft of its report. See "Report by the Special Committee of the National Security Council to the President," in *Foreign Relations of the United States, 1950,* vol. 1 (Washington, D.C., 1977), 513–23; Lilienthal, *Journals,* 2:623–33; Dean Acheson, *Present at the Creation: My Years in the State Department* (New York: Norton, 1969), 456–57.

15. Lilienthal, *Journals,* 3:5, unnumbered note. Lilienthal, in his *Journals,* mistakenly gives the impression that his first *Collier's* article appeared in 1950; actually, he

published his first article in that magazine a year earlier. See Lilienthal, "What Good is the Atom?" *Collier's,* June 11, 1949, 18–19, 55, 57; Lilienthal, "Free the Atom," *Collier's,* June 17, 1950, 13–15, 54, 56–58.

16. Lilienthal, *Journals,* 3:210–11. The Sherman Antitrust Act of 1890 (26 Stat. 209) forbade "every contact, combination . . . or conspiracy in the restraint of trade or commerce." Lilienthal presented a fuller treatment of his evolving views on business in his book, *Big Business: A New Era* (New York: Harper, 1953).

17. Lilienthal, *Journals,* 3:269; "Meeting Halfway," *New Yorker,* Dec. 8, 1951, 34–35. Helen was Lilienthal's wife.

Part V. A Return to Private Practice

1. For the report Swidler edited, see United Nations, *Integrated River Basin Development: A Report by a Panel of Experts* (New York: United Nations, Department of Economic and Social Affairs, 1958).

CHAPTER 14. KNOXVILLE AND NASHVILLE

1. John Corson (1905–1990) was the assistant executive director of the Social Security Board (1936–38), director of the Bureau of Old Age and Survivors Insurance (1938–41), an executive with the *Washington Post* (1945–50), and a management consultant for McKinsey and Company (1951–66).

2. Electric cooperatives, established for the purpose of providing at-cost services, are private utilities owned by consumers. The Joe Wheeler Electric Membership Cooperative was founded in 1937.

3. The litigation between Paducah and KU was lengthy and complex. In 1940, Paducah granted KU a twenty-year franchise to provide the city with electric power. That agreement contained a provision allowing the city to revoke the franchise, as long as it told KU of its intentions eighteen months in advance. The agreement also permitted the city to purchase KU's distribution system if both parties agreed on an acquisition price. In 1944, Paducah attempted to exercise both options; litigation resulted, but KU prevailed. What Swidler seems to have advised—although his explanation in the text is hard to parse—was for the city to forego renewed litigation and simply let the franchise expire. Were KU to continue objecting after this point, its reputation would have been tarnished, or so Swidler argued. In anticipation of the franchise's expiration, Swidler recommended that the city begin negotiating an acquisition price for KU's distribution system. Confronted by a public referendum authorizing purchase of the system—and by federal legislation allowing the TVA to extend service to Paducah—KU relented. On Aug. 31, 1961, KU delivered a release deed to its distribution system in exchange for $5,903,577.67. TVA service to Paducah was initiated on Jan. 1, 1963. See *Kentucky Utilities Co. v. City of Paducah,* 214 S.W.2d 258 (1948); *City of Paducah v. Kentucky Utilities,* 264 S.W.2d 848 (1953); "KU Delivers Deed of Release to City, Ending 20-Year Fight," *Paducah Sun Democrat,* Aug. 31, 1961, and "Paducah Gets KU System," *Louisville Courier-Journal,* Sept. 1, 1961, both located in box 10, folder 2, SP.

4. The Phillips Petroleum Corporation, incorporated in 1917, had revenues of approximately $12 billion by the late 1990s. The "percentage depletion" allowance

permitted producers to exempt a certain amount—eventually 27.5 percent—of their annual gross revenues from federal taxation (44 Stat. 9). This was based on the assumption that the depletion of a natural resource (for example, pumping oil out of a well and selling it) represented, for the owner of that resource, a so-called "wasting" capital asset. See Craig Goodman, "History of the U.S. Petroleum Depletion Allowance: 1890–1990," *Petroleum Accounting and Financial Management Journal* 11 (summer 1992): 120–50.

Part VI. Chairman of the Federal Power Commission

1. "Potomac Puzzle: Next FPC Chief," *NYT,* Jan. 23, 1962, sec. 3, 1.

2. "Extra-High-Voltage at the FPC," *Business Week,* Sept. 21, 1963, 80.

CHAPTER 15. BACK TO WASHINGTON, D.C.

1. Ralph Dungan (1923–) was a legislative assistant to Senator John Kennedy (1956–57), special assistant to President Kennedy (1961–63), special assistant to President Johnson (1963–64), and ambassador to Chile (1964–67). Adam Yarmolinsky (1922–2000) was a special assistant to the U.S. secretary of defense (1961–64), deputy director of the President's Task Force on Poverty (1964), and principal deputy assistant to the U.S. secretary of defense (1965–66).

2. For Kennedy's announcement, as well as coverage of Swidler's confirmation hearings, see "Kennedy Plans to Act Quickly to Check Decline in Economy," *NYT,* Jan. 27, 1961, 10; "Senate Confirms Swidler to FPC," ibid., June 15, 1961, 16; "FPC Chairman to Give Up Post," ibid., June 29, 1961, 29. Paul Rand Dixon (1913–1996) was a trial attorney (1938–57), chairman (1961–70), and member (1970–81) of the Federal Trade Commission. He was also chief counsel for the U.S. Senate Subcommittee on Antitrust and Monopoly (1957–61). Newton Minow (1926–) was an administrative assistant to Governor Adlai Stevenson (1952–53) and chairman of the FCC (1961–63). Adlai Stevenson (1900–1965) was special counsel to the Agricultural Adjustment Administration (1933–34), assistant to the secretary of the navy (1941–44), assistant to the U.S. secretary of state (1945), governor of Illinois (1949–53), U.S. representative to the United Nations (1960–65), and Democratic candidate for president of the United States in 1952 and 1956.

3. James Landis (1899–1964) was a professor at Harvard Law School (1926–33), a member of the FTC (1933–34), member (1934) and chairman (1935–37) of the Securities and Exchange Commission, dean of Harvard Law School (1937–46), director of the Office of Civil Defense (1942), and special assistant to President Kennedy (1961). Concerning the FPC, Landis concluded: "The Federal Power Commission without question represents the outstanding example in the federal government of the breakdown of the administrative process." See James M. Landis, *Report on Regulatory Agencies to the President-Elect, U.S. Senate, Committee on the Judiciary,* 86th Cong., 2nd sess. (Washington, D.C., 1960), 54. The six agencies to which Swidler is referring are the AEC (now the Nuclear Regulatory Commission), the FPC (now the Federal Energy Regulatory Commission), the Federal Communications Commission (FCC), the Federal Trade Commission (FTC), the Civil Aeronautics Board (CAB), and the Interstate Commerce Commission (ICC). The FCC

was created by the Communications Act of 1934 to regulate interstate and foreign communications by wire and radio (48 Stat. 1064). The FTC was established in 1914 by the Federal Trade Commission Act to prevent unfair or deceptive trade practices (38 Stat. 717). The CAB was created by the Civil Aeronautics Act in 1938 to promote and regulate the civil air industry in the U.S. and between the U.S. and foreign countries (52 Stat. 973). Many of the CAB's functions were transferred to the U.S. Commerce Department's National Transportation Safety Board in 1966; what remained of the CAB was terminated in 1984 (80 Stat. 931 and 98 Stat. 1703). The ICC was created by the Interstate Commerce Act in 1887 to regulate interstate surface transportation (24 Stat. 349). It was abolished by the ICC Termination Act of 1995 (109 Stat. 803); most of its functions were taken over by the Federal Highway Administration and the newly created Surface Transportation Board in the U.S. Department of Transportation.

4. Lawrence J. O'Connor Jr. (1914–1996) was assistant director of the Office of Oil and Gas in the U.S. Department of the Interior (1959–60), administrator of the Oil Import Administration (1960–61), and FPC commissioner (1961–71).

5. Andrew "Drew" Pearson (1897–1969) was one of America's most influential political columnists for more than thirty-five years. He established, along with fellow journalist Robert S. Allen (and, later, Jack Anderson), a syndicated column in 1932 that focused on exposing corruption and misconduct among public officials. Pearson's critics often accused him of distorted reporting and character assassinations.

6. Howard Morgan (1914–), a Democrat and advocate of public power systems, was a member of the Oregon legislature (1949–51), later serving as that state's sole public service commissioner (1957–59) and as an FPC commissioner (1961–63). Morgan apparently believed that Swidler and subsequent appointees to the FPC tended to side with private utilities in regulatory issues. While there is no evidence to support this claim, Morgan used it as the basis for declining his renomination to the FPC. His dissatisfaction with the FPC was detailed in a column written by Drew Pearson; that article likely confirmed Swidler's belief that Morgan had been the source of various press leaks. See "FPC Commissioner Spurns Second Term," *NYT,* Jan. 25, 1963, reprinted in full in *Congressional Record—Senate,* 88th Cong., 1st sess., Jan. 25, 1963, 1078–79.

7. Jerome Kuykendall (1907–1987) was an assistant U.S. attorney general (1941–44, 1946), chairman of the Washington Public Service Commission (1951–53), FPC chairman (1953–61), and member (1967–69) and chairman (1969–78) of the Indian Claims Commission.

8. Sewell Avery (1873–1960) was president (1905–37) and chairman of the board (1937–51) of the United States Gypsum Company and chairman of the board (1931–55) of Montgomery Ward and Company. Avery was removed from his office on Apr. 27, 1944. The conflict between Montgomery Ward and the federal government concerned the company's refusal to conform to directives issued by the War Labor Board. See "Biddle Gets Court Order to Restrain Ward Staff; Avery Evicted by Army," *NYT,* Apr. 28, 1944, 1, 14.

9. Lee White (1923–) was an attorney in the TVA's legal division (1950–54), legislative assistant to John F. Kennedy (1954–57), assistant special counsel to President

Kennedy (1961–63), associate (1963–65) and special counsel (1965–66) to President Johnson, and FPC chairman (1966–69).

10. Theodore Sorensen (1928–) was an assistant to Senator John Kennedy (1953–61) and special counsel to the president of the United States (1961–64).

11. Robert Kerr (1896–1963), a Democrat, was governor of Oklahoma (1943–47) and a U.S. senator from Oklahoma (1949–63).

12. The Edison Electric Institute (EEI) is a national trade association that provides information to, and lobbies for public policies on behalf of, shareholder-owned electric companies. Founded in 1933, the EEI was the successor organization to the National Electric Light Association, which was created in 1885.

13. Almer Monroney (1902–1980), a Democrat, was a U.S. representative (1939–51) and U.S. senator (1951–69) from Oklahoma.

14. Jacob Javits (1904–1986), a Republican, was a U.S. representative from New York (1947–54), attorney general of New York (1954–57), and U.S. senator from New York (1957–81).

15. John Pastore (1907–2000), a Democrat, served in the Rhode Island House of Representatives (1935–37), was lieutenant governor (1944) and governor (1945–50) of Rhode Island, and a U.S. senator from Rhode Island (1950–76). Warren Magnuson (1905–1989), a Democrat, was a member of the Washington House of Representatives (1933–34) and a U.S. representative (1937–44) and senator (1944–81) from Washington.

16. The Natural Gas Act of 1938 (52 Stat. 821) gave the FPC authority to regulate the interstate transmission and sale of natural gas. Initially, the FPC's regulation focused on the transmission industry—companies that specialized in building pipelines to convey natural gas from one point to another. In 1954, the Supreme Court ruled that the Natural Gas Act authorized the FPC to regulate the rates of companies—including independent producers—that were engaged in the interstate resale, not just transmission of, natural gas. The phrase "independent producer" was used to denote a company that was neither engaged in the interstate gas pipeline business, nor affiliated with an interstate gas pipeline company. See *Phillips Petroleum Co. v. State of Wisconsin*, 347 U.S. 672 (1954).

17. Harold Reis (1916–) served as an attorney in the U.S. Department of Justice (1941–64) and as executive assistant to the U.S. attorney general (1964–67).

CHAPTER 16. TURNING THE FPC AROUND

1. For a photo of the modified seating arrangement that Swidler used for commission meetings, see "Extra-High-Voltage at the FPC," 72–73.

2. Ex parte communication refers to any communication, oral or written, not on the public record. The FPC issued its prohibition against ex parte communications on Mar. 13, 1962. See Administrative Procedure Act, *U.S. Code* (1946), vol. 1, sec. 551; 60 Stat. 237; FPC, *Forty-First Annual Report of the Federal Power Commission, Fiscal Year Ended June 30, 1961* (Washington, D.C., 1962), 26.

3. Ralph Spritzer (1917–) was a solicitor in the U.S. Department of Justice (1946–61), general counsel of the FPC (1961–62), and senior assistant to the U.S. solicitor general (1962–68).

4. S. David Freeman (1926–), a lawyer and engineer for the TVA (1948–61), worked as Swidler's executive assistant at the FPC, served in the president's Office of Science and Technology (1967–71), was director of the Ford Foundation's Energy Policy Project (1971–74), general manager of the Lower Colorado River Authority (1986–90), general manager of the Sacramento Municipal Utility District (1990–94), and general manager of the Los Angeles Department of Water and Power (1997–). He was also a director (1977–78, 1981–84) and chairman (1978–81) of the TVA.

5. Alfred Kahn (1917–) is a noted economist who served in the anti-trust division of the U.S. Department of Justice (1941–42), the U.S. Department of Commerce (1942), and the War Production Board (1943); he was also a professor of economics at Cornell University (1947–89). Haskell Wald (1916–1981) was an economist for several federal agencies between 1938 and 1953, an economist for the Federal Reserve Bank of New York (1955–62), and chief economist for the FPC (1963–77). The FPC's economics division was established during the 1962 fiscal year. See FPC, *Forty-First Annual Report,* 27.

CHAPTER 17. THE DYNAMICS OF REGULATION UNDER THE NEW COMMISSION

1. Halberstam gave the following description of Swidler: "Joseph Swidler, chairman of the Federal Power Commission, a man strongly committed to regulating the big power and utility interest, found his first year with the Kennedy administration immensely frustrating. He had gone to Washington because he had been promised a strong anti-interest commission. That commission, he soon found out, would not be forthcoming. It was bogged down in the pluralism of American politics and by the president's primary concern with foreign affairs: In order to get his key foreign aid bills through Congress, Kennedy needed the cooperation of men like Sam Rayburn and Senator Robert Kerr. The price they extracted from the president was at the expense of the Federal Power Commission; they wanted and received men sympathetic to their and big interests' views. This left Swidler angry, and with the feeling that he was being betrayed by the administration." If Halberstam's portrayal is accurate, then Swidler's views changed over time. In short, he gives no indication in his memoir that he felt that his mission at the FPC had been comprised by having to accept certain nominees. Another explanation is possible. Halberstam, interested in making the point that Kennedy subordinated domestic concerns to foreign policy matters, might have overemphasized Swidler's discontent and frustration. See David Halberstam, *The Best and the Brightest* (New York: Random House, 1969), 126–27.

2. Maurine Neuberger (1907–2000), a Democrat, was a member of the Oregon House of Representatives (1951–55) and a U.S. senator from Oregon (1960–67).

3. David S. Black (1928–) was counsel for the Washington Public Service Commission (1957–61), general counsel for the U.S. Bureau of Public Roads (1961–63), a commissioner of the FPC (1963–66), administrator of the Bonneville Power Administration (1966–67), and undersecretary of the U.S. Department of the Interior (1967–69). Charles R. Ross (1920–) was chairman of the Vermont Public Service Commission (1959–61) and a commissioner of the FPC (1961–68). George Aiken (1892–1984), a Republican, was a member of the Vermont House of Representatives

(1931–35), lieutenant governor (1935–37) and governor (1937–41) of Vermont, and a U.S. senator from Vermont (1941–75). Henry "Scoop" Jackson (1912–1983), a Democrat, was a U.S. representative (1941–53) and a U.S. senator (1953–83) from Washington. He was an unsuccessful candidate for the Democratic nomination for president of the United States in 1972 and 1976.

4. Samuel Rayburn (1882–1961), a Democrat, was a member of the Texas House of Representatives (1907–13) and a U.S. representative from Texas (1913–61). He held the position of Speaker of the U.S. House of Representatives longer than any other individual (1940–47, 1949–53, 1955–61).

5. Harold C. Woodward (1901–1964) was a hearings referee for the Illinois Commerce Commission (1953–58) and a commissioner of the FPC (1962–64). Everett Dirksen (1896–1969) a Republican, was a U.S. representative (1933–39) and a U.S. senator (1951–69) from Illinois.

6. Carl Bagge (1927–) served as a special assistant (1963–64), general attorney (1964–65), and a commissioner (1965–70) of the FPC.

7. "Rulemaking" has emerged as an important aspect of administrative governance. A "rule" is a directive issued by a government agency. Rules carry the force of law, and are equivalent to statutes, executive orders, and judicial decisions. Rules can be related to enforcement of a particular statute, or they can focus more generally on assisting an agency to achieve its goals and objectives. The most explicit definition of rulemaking is outlined in the Administrative Procedure Act (1946). Supporters of limited government have become increasingly concerned over the growth and reach of rulemaking. See Cornelius M. Kerwin, *Rulemaking: How Government Agencies Write Law and Make Policy* (Washington, D.C.: Congressional Quarterly Press, 1994).

CHAPTER 18. HOW LIFE WAS BROUGHT TO FPC ELECTRIC POWER REGULATION

1. *Southern California Edison Company v. Federal Power Commission*, 310 F.2d 784 (1962); *Federal Power Commission v. Southern California Edison Company*, 376 U.S. 205 (1964).

2. Walker Cisler (1897–1994) was an electrical engineer who worked in the WPB (1941–43) and eventually became president (1951–64) and chairman of the board (1964–75) of the Detroit Edison Company.

3. James Couzens (1872–1936), a Republican, was mayor of Detroit (1919–22) and a U.S. senator from Michigan (1922–36).

4. The U.S. Department of Agriculture, through its Rural Electrification Administration (REA) and, later, through its Rural Utilities Service, was authorized to make low-interest loans to electric cooperatives. See 49 Stat. 1363; 108 Stat. 3219.

CHAPTER 19. NEW PROGRAMS IN ELECTRIC POWER REGULATION AND THE NORTHEAST POWER FAILURE

1. A description of Swidler's recreation initiative, which was begun during the 1963 fiscal year, can be found in FPC, *Forty-Fifth Annual Report of the Federal Power Commission, Fiscal Year Ended June 30, 1965* (Washington, D.C., 1965), 71.

2. Swidler is likely referring to the Vaiont Dam disaster, which occurred on Oct. 9, 1963. It resulted in more than 2,500 deaths.

3. U.S. House, *Hearings before a Subcommittee of the Committee on Appropriations (Independent Offices Appropriations for 1964)*, 88th Cong., 1st sess., Jan. 28, 1963, quotes at 365. Section 10(e) of the Federal Water Power Act is located at 41 Stat. 1069.

4. The REA was established on May 11, 1935, by Executive Order 7037. Initially placed in the U.S. Department of the Interior, then moved to the U.S. Department of Agriculture in 1939, the purpose of the REA was to extend electrification throughout rural America by making low-interest loans available to electrical cooperatives. The REA's functions were taken over by the Rural Utilities Service in 1994. See 108 Stat. 3219. Two of the cases that were used to test the possibility of FPC jurisdiction over electric cooperatives were *South Central Rural Electric Cooperative v. Swidler*, 228 F.Supp. 875 (1963), and *Minnkota Power Cooperative v. Swidler*, 228 F.Supp. 968 (1963).

5. The American Public Power Association, founded in 1940, is a national trade association representing more than two thousand municipal and other government-owned electric utilities.

6. See U.S. Senate, *Hearings before the Committee on Commerce on S. 2028 and S. 3038 (Clarifying Power Commission Jurisdiction)*, 88th Cong., 2nd sess., July 22, 1964, 16–47, especially 22, 26, 36. Norman M. Clapp (1914–1998) was a mediator for the Wisconsin Labor Relations Board (1937–39), administrator of the Rural Electrification Administration (1961–69), and chairman of the Wisconsin Public Service Commission (1975–77). Thurmond announced his switch to the Republican Party on Sept. 16, 1964.

7. On Jan. 5, 1967, the FPC concluded that it did not have jurisdiction over generation and transmission cooperatives. The U.S. Court of Appeals for the District of Columbia affirmed this ruling on Feb. 15, 1968. See *Dairyland Power Cooperative et al.*, 37 FPC 12 (1967); *Salt River Project Agricultural Improvement and Power District v. FPC*, 391 F.2d 470 (1968).

8. Buford Ellington (1907–1972), a Democrat, was a member of the Tennessee General Assembly (1949–51), agriculture commissioner of Tennessee (1953–58), governor of Tennessee (1959–63, 1967–71), and director of the U.S. Office of Emergency Planning (1965–66). A summary of, and direct quotation from, Johnson's memorandum to Swidler can be found in "U.S. Orders an Inquiry," *NYT*, Nov. 10, 1965, 1, 3.

9. Swidler's announcement was made on Nov. 15, almost a week after the blackout. For the text of his statement, as well as commentary on the reasons behind delaying the FPC's initial findings, see "Preliminary Report Warns It Could Happen Again," *NYT*, Nov. 16, 1965, 1, 58. Excerpts from the FPC's final report, dated Dec. 6, 1965, were reprinted in "FPC Criticizes Power Systems in Nov. 9 Failure," *NYT*, Dec. 7, 1965, 1, 40.

CHAPTER 20. THE NATIONAL POWER SURVEY

1. FPC, *National Power Survey: A Report by the Federal Power Commission* (Washington, D.C., 1964).

2. Stewart Udall (1920–), a Democrat, served as a U.S. representative from Arizona (1955–61) and U.S. secretary of the interior (1961–69).

3. For the establishment of the FPC's advisory committees, see FPC, *Forty-First Annual Report,* 22.

4. Section 202(a) is located at 49 Stat. 848.

5. A copy of Swidler's speech was reprinted in the *Edison Electric Institute Bulletin* 31 (June–July, 1963): 200–204, 245.

6. The Electric Research Council (ERC) was established in 1965. The Electric Power Research Institute, created in 1973, took over many of the research and development projects previously supervised by the ERC and the Edison Electric Institute.

7. Hoping to prevent another major blackout, electric utilities created the National Electric Reliability Council (now the North American Electric Reliability Council) in 1968. NERC is a nonprofit organization comprised of ten regional councils, whose members represent all segments of the electric utilities industry.

8. U.S. House, *Hearings before a Subcommittee of the Committee on Appropriations (Independent Offices Appropriations for 1966),* 89th Cong, 1st sess., Mar. 29, 1965, 551.

9. FPC, *1970 National Power Survey of the Federal Power Commission* (Washington, D.C., 1970). John N. Nassikas (1917–1998), a Republican, was assistant deputy attorney general for New Hampshire (1950–53), FPC chairman (1969–75), and a member of the New Hampshire Public Service Commission (1984–86).

CHAPTER 21. THE NATURAL GAS PROGRAM

1. The Supreme Court case Swidler is referring to is *Phillips Petroleum Co. v. State of Wisconsin,* 347 U.S. 672 (1954). See also n. 3 below.

2. In the spring of 1961, President Kennedy lamented that there were more than four thousand rate cases pending before the FPC. Among other recommendations, Kennedy suggested—without success—that the FPC be given two additional commissioners. See John F. Kennedy, "Special Message to the Congress on Regulatory Agencies," Apr. 13, 1961, in *Public Papers of the Presidents of the United States: John F. Kennedy, 1961–1963,* vol. 1 (Washington, D.C., 1962), 272–73.

3. In the *Phillips* decision (1954), the Supreme Court ruled that the FPC could regulate the resale prices of independent natural gas producers engaged in interstate commerce, but did not specify how the FPC was supposed to determine those rates. Rather than hold rate hearings on each of the more than five thousand producers who were involved in interstate gas transactions, the FPC decided to establish an "area rate" for each of the country's gas producing regions. Between 1960 and 1965, the FPC used an interim area-rate schedule called the "guideline price policy," which was eventually ruled constitutional by the courts. See *Phillips Petroleum Company,* 10 FPC 246 (1951); *Wisconsin v. Federal Power Commission,* 373 U.S. 294 (1963). In August 1965, the FPC revised its area-rate schedule, starting first with the region known as the Texas–New Mexico Permian Basin. With minor modifications, the FPC's new rates were held constitutional. See *Permian Basin Area Rate Cases,* 390 U.S. 747 (1968). See also *Forty-Sixth Annual Report of the Federal Power Commission, July 1, 1965–June 30, 1966* (Washington, D.C., 1967), 128–30; M. Elizabeth Sanders, *The Regulation of Natural Gas: Policy and Politics, 1938–1978* (Philadelphia: Temple Univ. Press, 1981), 96, 111–13.

4. According to *U.S. Government Manuals,* John C. Mason replaced Malloy J. McQueen as the FPC's deputy general counsel during the early phase of Swidler's tenure. It is unclear, though, if this is the individual who decided to resign rather than implement the new settlement program.

5. The FPC referred to the expedited decision procedure as the "abridged hearing procedure." For a description of this process, see FPC, *Forty-First Annual Report,* 89, 125.

6. *Tennessee Gas Transmission Company v. Federal Power Commission,* 340 F.2d 100 (1964); *Tennessee Gas Transmission Company v. Federal Power Commission,* 381 U.S. 950 (1965).

7. William Elmer (1915–) was vice-president (1946–57), president (1957–68), and chairman of the board (1968–80) of the Texas Gas Transmission Corporation. For establishment of the Natural Gas Advisory Council, see FPC, *Forty-First Annual Report,* 21.

CHAPTER 22. CONGRESSIONAL CONFLICTS: ALBERT THOMAS AND OTHER REPRESENTATIVES

1. Albert Thomas (1898–1966) was an assistant U.S. district attorney for the southern district of Texas (1930–36) and a U.S. representative from Texas (1937–66).

2. Lawrence F. O'Brien (1917–1990) was special assistant for congressional relations to Presidents Kennedy and Johnson (1961–65) and chairman of the Democratic National Committee (1968–69, 1970–72). Walter Heller (1915–1987) was an economist and consultant to the U.S. Department of Treasury (1942–53), professor of economics at the University of Minnesota (1946–86), and chairman of the Council of Economic Advisers to the President (1961–64).

3. U.S. House, *Hearings before a Subcommittee of the Committee on Appropriations (Independent Offices Appropriations for 1965),* 88th Cong., 2nd sess., Feb. 19, 1964, quotes at 806, 807. Joseph L. Evins (1910–1984), a Democrat, was a counsel (1935–38) and assistant secretary (1938–40) for the FTC, and a U.S. representative from Tennessee (1947–77).

4. Charles Buckley (1890–1967), a Democrat, was a member of the New York City Board of Aldermen (1918–23), New York State tax appraiser (1923–29), chamberlain of New York City (1929–33), a U.S. representative from New York (1935–65) and chairman of the House Public Works Committee (1951–52, 1955–64).

5. Swidler does not identify the member of Congress at whose home the party occurred. Russell Long (1918–), a Democrat and son of Huey P. Long (1893–1935), was a U.S. senator from Louisiana (1948–86) and chairman of the Senate Finance Committee (1965–80).

6. Richard Ottinger (1929–), a Democrat, served as a director of programs for the Peace Corps (1961–64) and was a U.S. representative from New York (1965–71, 1975–85).

7. What Swidler likely meant by "political experience" was elected office.

CHAPTER 23. OFFICIAL TRIPS TO THE U.S.S.R. AND LONDON

1. James K. Carr (1914–1980), a civil engineer, worked in the U.S. Bureau of Reclamation (1936–51), was an engineering consultant to the House Committee on

Interior and Insular Affairs (1951–53), assistant general manager of the Sacramento Municipal Utility District (1953–61), undersecretary of the U.S. Department of the Interior (1961–64), general manager of public utilities for San Francisco (1964–70), and director of the San Francisco International Airport (1970–73). Aubrey Wagner (1912–1990) was an engineering aid (1934–51), assistant general manager (1951–61), director (1961–62), and chairman (1962–78) for the TVA. Fred Chambers (1912–1993), an electrical engineer, worked for several utility companies prior to joining the TVA (1939–71); he also served on the New York State Public Service Commission (1973–74). Charles F. Luce (1917–) was the administrator of the Bonneville Power Administration (1961–66), undersecretary of the U.S. Department of the Interior (1966–67), and chief executive officer of Consolidated Edison of New York (1967–81). Tellingly, Swidler omitted the fact that Howard Morgan was also a member of the delegation. For the other members, see "Recent Electric Power Developments in the U.S.S.R.," Dec. 27, 1962, 3–4, in container 13, folder 3, SP. The trip lasted from Aug. 28–Sept. 9, 1962.

2. D. N. Poliakov was officially listed in the trip itinerary as a representative of the State Committee for Coordination of Scientific Works. The woman to whom Swidler refers was likely Valentina M. Gusarova, a Soviet interpreter. A copy of the itinerary can be found in container 13, folder 3, SP.

3. Nikita Sergeyevich Khrushchev (1894–1971) was the first secretary of the Soviet Communist Party (1953–64), as well as leader (1956–58) and premier (1958–64) of the Soviet Union. The person Swidler identified as the minister of energy was probably Pjotr S. Neporozhny, the Deputy Minister of Power Station Construction; the actual minister of power, I. T. Novikov, was absent. Neporozhny became the minister of power and electrification in November of 1962. After returning from the trip, Swidler learned from protocol officers in the U.S. State Department that Carr, not he, should have been second in command.

4. Lake Baikal, located in south central Siberia, is the deepest freshwater lake in the world. The answers to Swidler's questions were published in *Electrical World* 158 (Dec. 17, 1962): 45.

5. "Potemkin Village" refers to a pretentiously showy façade intended to mask or divert attention from an embarrassing or shabby condition. Irkutsk, located on the Angara River, is one of the major industrial cities in Siberia.

6. The Bratsk Dam is on the Angara River in south central Siberia. It is 410 feet high and 14,488 wide at the crest. It has an electric power capacity of 4,500 megawatts. The Angara River is the only outlet for Lake Baikal. Perhaps the dam's "Potemkin Village" quality was due to the fact that it was not completed until 1964.

7. The incident Swidler is referring to is the so-called "U-2 affair." Gary Powers (1929–1977), an American air force pilot working for the Central Intelligence Agency, was shot down over Soviet territory on May 1, 1960. Captured and interrogated, Powers spent two years in prison before being released in exchange for Soviet spy Rudolf Abel. When the United States refused to apologize for the U-2 affair, Khrushchev withdrew a previously extended invitation for Eisenhower to visit the U.S.S.R. The designation "U-2" refers to the type of plane Powers was piloting. See Nikita Sergeyevich Khrushchev, *Khrushchev Remembers: The Last Testament*, trans.

and ed. Strobe Talbott, with a foreword by Edward Crankshaw and an introduction by Jerrold L. Schecter (Boston: Little Brown, 1974), 452.

8. Manuscript and typescript versions of Swidler's diary are in container 13, folder 2, SP.

9. Fyodor Dostoyevsky (1821–1881) was a Russian novelist whose works embraced the social and political issues of nineteenth-century Russia.

10. Model, Roland, and Company, a New York–based investment bank founded in 1941, merged with Shields and Company in 1974; the Bache securities group acquired Shields, Model, Roland in 1977. The House of Rothschild is a powerful, international consortium of banks founded by Mayer Anschel Rothschild (1744–1822) in Germany during the late eighteenth century. In his speech, Swidler emphasized the inability of regulatory agencies to confiscate property or prevent a fair rate of return. See "The Program and Activities of the Federal Power Commission," speech, Dec. 9, 1963, container 18, folder 14, SP.

11. A turnkey contract is an agreement whereby a vendor, usually after winning a competitive bidding process, agrees to supply a ready-to-operate installation (which is often publicly owned).

CHAPTER 24. LEAVING THE FPC

1. John W. Macy Jr. (1917–1986) was the executive director (1953–58) and chairman (1961–69) of the U.S. Civil Service Commission, president of the Corporation for Public Broadcasting (1969–72), and director of the Federal Emergency Management Agency (1979–81).

2. Bill D. Moyers (1934–), a journalist and television commentator, was associate director (1961–62) and deputy director (1963) of the Peace Corps. He also served as a special assistant (1963–67) and press secretary (1965–67) for President Johnson.

3. S. David Freeman, *Energy: The New Era* (New York: Walker, 1974), 186–88. The passage that Swidler refers to is on page 186: "Joseph Swidler, who was appointed by President Kennedy, was not renominated by President Johnson. Instead Johnson appointed another consumer man, Lee White, his White House counsel, who had not yet incurred the wrath of industry."

4. Glenn T. Seaborg (1912–1999), recipient of the Nobel Prize in chemistry (1951), served on the AEC's general advisory committee (1946–50) and as the AEC's chairman (1961–71). James T. Ramey (1914–) was an attorney for the TVA (1941–47), an attorney for the AEC (1947–56), executive director of the Joint Congressional Committee on Atomic Energy (1956–62), and member of the AEC (1962–73).

Part VII. Back to Private Practice

CHAPTER 25. A LAW PRACTICE INTERLUDE IN WASHINGTON, D.C.

1. Richard M. Freeman (1921–) was an attorney (1948–57) and director (1978–86) of the TVA.

2. William I. Harkaway (1928–) worked as an attorney in the U.S. Department of Labor (1956–58) and as principal staff attorney for the FPC (1958–67).

3. Fredrick M. Eaton (1905–1984) was a corporate lawyer and a partner in the firm of Shearman and Sterling.

4. John V. Lindsay (1921–2000) was executive assistant to the U.S. attorney general (1955–57), a U.S. representative from New York (1959–65), and mayor of New York City (1965–73); he changed party affiliation from Republican to Democrat in 1971. Charles F. Luce, *Some Lessons Learned: Recollections of 15 Years as Chairman of Consolidated Edison, 1967–1982* (New York: Consolidated Edison of New York, 1990), 30.

5. For the fullest treatment of the paving scandal, see Walter Goodman, *A Percentage of the Take: A Classic Case of Big City Corruption* (New York: Farrar, Straus, and Giroux, 1971).

Part VIII. Chairman of the New York State Public Service Commission

1. For complaints about the PSC, see "PSC Criticized on Utility Costs," *NYT,* July 1, 1970, 20.

2. NYPSC, *Annual Report of the New York State Public Service Commission for the Year Ending December 31, 1970* (Albany, n.d.), quote at 66. Administratively, the New York legislature addressed the issue of pollution by creating a Department of Environmental Conservation on Apr. 23, 1970. This consolidated the state's three antipollution agencies within one department. See "Pollution Unit Gains in Albany," *NYT,* Apr. 19, 1970, 1; "Activity Ranges from Oratory to Legislation," ibid., Apr. 23, 1970, 30.

3. OPEC, a petroleum cartel, was formed on Sept. 14, 1960, by Iran, Iraq, Kuwait, Saudi Arabia, and Venezuela. Subsequent members included Qatar (1961), Indonesia and Libya (1962), Abu Dhabi (1967), Algeria (1969), Nigeria (1971), and Ecuador and Gabon (1973). Ecuador withdrew from the organization on Jan. 1, 1993. By the mid-1990s, OPEC accounted for approximately 43 percent of the world's production of petroleum. PSC data come from NYPSC, *Annual Report, 1970,* 5.

4. NYPSC, *Annual Report of the New York State Public Service Commission for the Year Ending December 31, 1971* (Albany, n.d.), quote at 37.

CHAPTER 26. WORKING FOR ROCKEFELLER IN ALBANY

1. Harry W. Albright Jr. (1925–) held various positions in the New York State governor's office (1967–72), was New York State bank superintendent (1972–74), and served as special counsel to Vice-President Nelson Rockefeller (1974–75). Nelson A. Rockefeller (1908–1979), a Republican and son of John D. Rockefeller Jr., served as an assistant U.S. secretary of state (1944–45), undersecretary of the U.S. Department of Health, Education, and Welfare (1953–54), governor of New York (1958–73), and vice-president (1974–77).

2. Swidler served as chairman of the PSC from Feb. 1, 1970, until May 30, 1974. He was reappointed on Apr. 18, 1974, to a second six-year term, but chose to resign instead. See NYPSC, *Annual Report, Year Ended December 31, 1974* (Albany, n.d.), 5.

3. According to its annual report for 1969, the PSC, as mandated by 1970 legislation, had to hold hearings on "rate increase proposals of utilities when they would increase annual revenues by more than $100,000 or by more than 2.5 percent of company revenues." See NYPSC, *Annual Report of the Public Service Commission, Department of Public Service, for the Year 1969* (Albany, n.d.), 3.

4. Dennis A. Rapp (1929–) was a program analyst for the U.S. Bureau of Land Management (1956–59), a senior examiner in the U.S. Bureau of the Budget (1961–65), and director of the New York State Office of Environmental Planning (1970–73). For a description of the PSC's Office of Environmental Planning, see NYPSC, *Annual Report, 1971,* 42–48.

5. Alvin Kaufman (1928–) was a mineral economist for the Alaska Bureau of Mines (1956–60), the U.S. Bureau of Mines (1960–67, 1969–70), and the U.S. Bureau of Land Management (1967–69). He worked at the PSC from 1970 until 1976. For a description of the PSC's Office of Economic Research, see NYPSC, *Annual Report, 1971,* 49–53. NARUC was founded in 1889. It is a nonprofit organization composed of public agencies that regulate telecommunications, energy, and water. NARUC established the NRRI at Ohio State University in 1976; it conducts research in utility-related issues.

6. Carmel Carrington Marr (1921–) was a legal adviser to the U.S. Mission to the United Nations (1953–67), senior legal officer to the United Nations Secretariat (1967–68), member of the New York State Human Rights Board (1968–71), and a commissioner of the PSC (1971–86).

7. Jurisdiction over surface transportation was transferred to the New York State Department of Transportation on Mar. 1, 1971. See NYPSC, *Annual Report, 1969,* 1.

CHAPTER 27. REGULATION IN AN INFLATIONARY ENVIRONMENT

1. Though Swidler defines "regulatory compact" to include the telephone industry, that particular concept usually refers to the electric power industry. In brief, the regulatory compact describes a set of mutual rights and obligations between the public (as represented by regulatory commissions) and private utilities—most commonly, electric companies. In exchange for receiving an exclusive franchise (that is, a legal monopoly) to provide service to a particular area, a utility agrees to minimize costs and maximize efficiency. Upon determining that a utility has met these criteria, a regulatory commission establishes rates that guarantee a reasonable rate of return (profit) for the company in question. Thus, supporters of the regulatory compact argue that it insures high levels of service at low costs, but without alienating private sector involvement and investment.

2. Edward P. Larkin (1915–1986), a Republican from Hemstead, New York, served in the New York State Assembly (1953–55) and Senate (1956–61), before being appointed to the PSC (1961–86). The PSC initially approved a $136.7 million rate increase on Feb. 26, 1970, for New York Telephone; it subsequently reduced that provisionally approved increase to $120.7 million, to become effective on Sept. 1, 1970. See NYPSC, *Annual Report, 1970,* 35.

3. A "tariff rate" usually refers to the monthly charge, excluding taxes and other fees, levied by a local telephone company.

4. The PSC adopted the Basic Service Index Plan on Jan. 17, 1972. See NYPSC, *Annual Report, 1971,* 26; NYPSC, *Annual Report, Year Ended December 31, 1972* (Albany, n.d.), 41; NYPSC, *Annual Report, Year Ended December 31, 1973* (Albany, n.d.), 45.

5. The PSC ordered New York Telephone to investigate ways to improve the quality of its coin-operated telephones on June 3, 1970. See NYPSC, *Annual Report, 1970,* 32.

6. The PSC permitted hotels to charge rates of fourteen cents for local calls, and between twenty and thirty cents for intrastate toll calls placed through a hotel's facilities. See NYPSC, *Annual Report, 1972,* 45.

7. For details concerning the supply of electricity during the summer of 1970, see NYPSC, *Annual Report, 1970,* 8–11.

8. There is no record of a "John Anderson" in the New York Senate while Swidler was PSC chairman. It seems likely that he meant Warren M. Anderson (1915–), a Republican member of the New York Senate (1953–88); Anderson served as chairman of the finance committee (1966–72) and was president pro tempore (1973–88).

9. Swidler was probably referring to Stanley Steingut (1920–1989), a Brooklyn Democrat who served in the New York State Assembly for twenty-six years (1952–78); he was minority leader from 1969 until 1974. Irving Like (1924–), a longtime environmental lawyer and activist, helped create the Fire Island National Seashore. He also successfully opposed several attempts by the Long Island Lighting Company to build nuclear plants on Long Island.

10. The bill discussed by Swidler was an amendment to the New York State Public Service Law. The legislature passed it on Apr. 29, 1970, and it became effective on July 1, 1970. See *Laws of New York,* 1970, chapter 272, 1524–35.

Part IX. Final Years

CHAPTER 28. A SHORT APOLOGIA FOR UTILITY REGULATION

1. In response to the energy crisis of the 1970s, federal legislation (namely, the Public Utility Regulatory Policies Act of 1978 [PURPA]) stimulated a new class of power producers. This segment of the energy industry now includes a variety of different entities, including "independent power producers," "nonutility generators," and "co-generators." Though not all of these producers trace their development from PURPA, they share some important attributes: They are not utilities in the traditional sense of the word, nor do they fall under the jurisdiction of regulatory commissions. Their primary function is to sell power to regulated utilities; the resulting competition supposedly lowers the end cost to consumers. For Swidler's views on these developments, see Swidler, "A Brave New World: Let's Look before We Leap," *Electricity Journal* 30 (Nov. 1990): 40–43; Swidler, "The Role of Competition in Bulk Power Supply," speech, North American Electric Reliability Council, general meeting, June 5, 1991, unorganized 1999 addition to SP.

Bibliography

Manuscript Collections

Booth, Harry R. Papers. Chicago Historical Society.
Frankfurter, Felix. Papers. Library of Congress.
Lilienthal, David E. Papers. Princeton University.
An Oral History of the Tennessee Valley Authority. University of Memphis.
Swidler, Joseph C. Papers. Library of Congress.
Tennessee Valley Authority Records and Files. National Archives and Records Administration, Southeast Region, East Point, Georgia.

Newspapers

Chicago Herald and Examiner.
Chicago Daily Tribune.
New York Times.

Joseph Swidler: Select Publications

"Rate Regulation in California: Some Significant Problems Raised by the California Railroad Commission." *Journal of Land & Public Utility Economics* 7 (Feb. 1931): 1–12.
"Utility Rate-Making in the Depression." *Journal of Land & Public Utility Economics* 8 (Nov. 1932): 347–51.
"TVA in Court: A Study of TVA's Constitutional Litigation." *Iowa Law Review* 32 (Jan. 1947): 296–326. (Coauthored with Robert H. Marquis.)
"A Brave New World: Let's Look before We Leap." *Electricity Journal* 30 (Nov. 1990): 40–43.
"An Unthinkably Horrible Situation," *Public Utilities Fortnightly* 128 (Sept. 15, 1991): 14–18, 40.

Books

Acheson, Dean. *Present at the Creation: My Years in the State Department.* New York: Norton, 1969.
Bonbright, James C., and Gardiner C. Means. *The Holding Company: Its Public Significance and Its Regulation.* New York: McGraw-Hill, 1932.
Breyer, Stephen G., and Paul W. MacAvoy. *Energy Regulation by the Federal Power Commission.* Washington, D.C.: Brookings Institution, 1974.
Clapp, Gordon R. *The TVA: An Approach to the Development of a Region.* Chicago: Univ. of Chicago Press, 1955.
Clarke, Jeanne Nienaber. *Roosevelt's Warrior: Harold L. Ickes and the New Deal.* Baltimore: Johns Hopkins Univ. Press, 1996.
Freeman, S. David. *Energy: The New Era.* New York: Walker, 1974.
Funigiello, Phillip J. *Toward a National Power Policy: The New Deal and the Electric Utility Industry, 1933–1941.* Pittsburgh: Univ. of Pittsburgh Press, 1973.

Halberstam, David. *The Best and the Brightest.* New York: Random House, 1969.
Hargrove, Erwin C. *Prisoners of Myth: The Leadership of the Tennessee Valley Authority, 1933–1990.* Princeton: Princeton Univ. Press, 1994.
Hargrove, Erwin C., and Paul K. Conkin, eds. *TVA: Fifty Years of Grass-Roots Bureaucracy.* Chicago: Univ. of Illinois Press, 1983.
Hawley, Ellis W. *The New Deal and the Problem of Monopoly.* Princeton: Princeton Univ. Press, 1966.
Hobday, Victor C. *Sparks at the Grassroots: Municipal Distribution of TVA Electricity in Tennessee.* Knoxville: Univ. of Tennessee Press, 1969.
Hubbard, Preston J. *Origins of the TVA: The Muscle Shoals Controversy, 1920–1932.* Nashville: Vanderbilt Univ. Press, 1961.
Ickes, Harold L. *The Secret Diary of Harold L. Ickes: The First Thousand Days.* New York: Simon and Schuster, 1953.
Kerwin, Cornelius M. *Rulemaking: How Government Agencies Write Law and Make Policy.* Washington, D.C.: Congressional Quarterly Press, 1994.
Lear, Linda J. *Harold L. Ickes: The Aggressive Progressive, 1874–1933.* New York: Garland, 1981.
Leuchtenburg, William E. *The FDR Years: On Roosevelt and His Legacy.* New York: Columbia Univ. Press, 1995.
Lilienthal, David E. *TVA: Democracy on the March.* Rev. ed. New York: Harper, 1953.
———. *Big Business: A New Era.* New York: Harper, 1953.
———. *The Journals of David E. Lilienthal.* 7 vols. New York: Harper, 1964–83.
Lowitt, Richard. *George W. Norris: The Triumph of a Progressive, 1933–1944.* Urbana: Univ. of Illinois Press, 1978.
Luce, Charles F. *Some Lessons Learned: Recollections of 15 Years as Chairman of Consolidated Edison, 1967–1982.* New York: Consolidated Edison of New York, 1990.
McCraw, Thomas. *Morgan vs. Lilienthal: The Feud within the TVA.* Chicago: Loyola Univ. Press, 1970.
———. *TVA and the Power Fight, 1933–1939.* Philadelphia: J. P. Lippincott, 1971.
———. *Prophets of Regulation: Charles Francis Adams, Louis D. Brandeis, James M. Landis, Alfred E. Kahn.* Cambridge: Harvard Univ. Press, 1984.
McDonald, Forrest. *Insull.* Chicago: Univ. of Chicago Press, 1962.
Morgan, Arthur E. *The Making of the TVA.* Buffalo: Prometheus Books, 1974.
Neuse, Steven M. *David E. Lilienthal: The Journey of an American Liberal.* Knoxville: Univ. of Tennessee Press, 1996.
Nye, David E. *Electrifying America: Social Meanings of a New Technology.* Cambridge: MIT Press, 1990.
———. *Consuming Power: A Social History of American Energies.* Cambridge: MIT Press, 1998.
Owen, Marguerite. *The Tennessee Valley Authority.* New York: Praeger, 1973.
Platt, Harold L. *The Electric City: Energy and the Growth of the Chicago Area, 1880–1930.* Chicago: Univ. of Chicago Press, 1991.
Pritchett, C. Herman. *The Tennessee Valley Authority: A Study in Public Administration.* Chapel Hill: Univ. of North Carolina Press, 1943.
Public Ownership League of America. *Proceedings: Public Ownership Conference, November 15–17, 1919.* Chicago: Public Ownership League of America, 1919.

Rose, Mark H. *Cities of Light and Heat: Domesticating Gas and Electricity in Urban America.* University Park: Pennsylvania State Univ. Press, 1995.

Sanders, Elizabeth M. *The Regulation of Natural Gas: Policy and Politics, 1938–1978.* Philadelphia: Temple Univ. Press, 1981.

Schwarz, Jordan A. *The New Dealers: Power Politics in the Age of Roosevelt.* New York: Random House, 1993.

Selznick, Philip. *TVA and the Grass Roots: A Study in the Sociology of Formal Organization.* Berkeley: Univ. of California Press, 1949.

Stern, Philip M. *The Oppenheimer Case: Security on Trial.* New York: Harper, 1969.

Talbert, Roy, Jr. *FDR's Utopian: Arthur Morgan of the TVA.* Jackson: Univ. Press of Mississippi, 1987.

Twentieth Century Fund. *Electric Power and Government Policy: A Survey of the Relations between the Government and the Electric Power Industry.* New York: Twentieth Century Fund, 1948.

Waldavsky, Aaron. *Dixon-Yates: A Study in Power Politics.* New Haven: Yale Univ. Press, 1962.

Willkie, Wendell. *This Is Wendell Willkie: A Collection of Speeches and Writings on Present-Day Issues.* New York: Dodd, Mead, 1940.

Articles

"Extra-High-Voltage at the FPC." *Business Week,* Sept. 21, 1963, 80.

Goodman, Craig. "History of the U.S. Petroleum Depletion Allowance: 1890–1990." *Petroleum Accounting and Financial Management Journal* 11 (summer 1992): 120–50.

Haimbaugh, George D., Jr. "The TVA Cases: A Quarter Century Later." *Indiana Law Journal* 41 (winter 1966): 197–227.

Lilienthal, David E. "What Good is the Atom?" *Collier's,* June 11, 1949, 18–19, 55, 57.

———. "Free the Atom." *Collier's,* June 17, 1950, 13–15, 54, 56–58.

Government Publications

Federal Power Commission. *Annual Reports.* 1960–66.

———. *National Power Survey: A Report by the Federal Power Commission.* Washington, D.C.: Government Printing Office, 1964.

Illinois Commerce Commission. *Opinions and Orders.* 1930–35.

Landis, James M. *Report on Regulatory Agencies to the President-Elect.* Washington, D.C.: Government Printing Office, 1960.

New York State Public Service Commission. *Annual Reports.* 1969–75.

Public Papers of the Presidents of the United States: Dwight D. Eisenhower, 1953–1960. 8 vols. Washington, D.C.: Government Printing Office, 1954.

Public Papers of the Presidents of the United States: John F. Kennedy, 1961–1963. 3 vols. Washington, D.C.: Government Printing Office, 1962.

Tennessee Valley Authority. *Annual Reports.* 1934–56.

U.S. Congress. *Hearings before the Joint Committee on the Investigation of the Tennessee Valley Authority.* 75th Cong., 3rd sess., 1938.

U.S. Congress. House. *Hearings before a Subcommittee of the Committee on Appropriations (Independent Offices Appropriations for 1965).* 88th Cong., 2nd sess., 1964.

U.S. Congress. House. *Hearings before a Subcommittee of the Committee on Appropriations (Independent Offices Appropriations for 1966)*. 89th Cong, 1st sess., 1965.

U.S. Congress. House. *Hearings before a Subcommittee of the Committee on Appropriations (Independent Offices Appropriations for 1964)*. 88th Cong., 1st sess., 1963.

U.S. Congress. Senate. *Hearings before the Committee on Commerce on S. 2028 and S. 3038 (Clarifying Power Commission Jurisdiction)*. 88th Cong., 2nd sess., 1964.

U.S. Congress. Senate. *Hearings before the Committee on Interstate and Foreign Commerce (Nomination of Joseph C. Swidler to be a Member of the Federal Power Commission)*. 87th Cong., 1st sess., 1961.

Federal Legislation

41 Stat. 1063. "Federal Water Power Act." Public Law 66–412 (1920).

46 Stat. 797. "Federal Power Act of 1930." Public Law 71–280 (1930).

48 Stat. 58. "Tennessee Valley Authority Act." Public Law 73–17 (1933).

48 Stat. 195. "National Industrial Recovery Act." Public Law 73–67 (1933).

49 Stat. 803. "Public Utility Holding Company Act of 1935." Public Law 74–333 (1935).

49 Stat. 1076. "An Act to Amend the Tennessee Valley Authority Act." Public Law 74–412 (1935).

52 Stat. 821. "The Natural Gas Act of 1938." Public Law 75–688 (1938).

73 Stat. 280. "An Act to Amend the Tennessee Valley Authority Act, As Amended." Public Law 86–137 (1959).

91 Stat. 565. "U.S. Department of Energy Organization Act." Public Law 95–91 (1977).

92 Stat. 3117. "Public Utility Regulatory Policies Act of 1978." Public Law 95–617 (1978).

Court Cases

Ashwander v. TVA, 297 U.S. 288 (1936).

Tennessee Electric Power Company v. Tennessee Valley Authority, 21 F.Supp. 947 (1938).

Alabama Power Company v. Ickes, 302 U.S. 464 (1938).

Tennessee Electric Power Company v. TVA, 306 U.S. 118 (1939).

Kentucky Utilities Co. v. City of Paducah, 214 S.W.2d 258 (1948).

City of Paducah v. Kentucky Utilities, 264 S.W.2d 848 (1953).

Phillips Petroleum Co. v. State of Wisconsin, 347 U.S. 672 (1954).

Southern California Edison Company v. Federal Power Commission, 310 F.2d 784 (1962).

South Central Rural Electric Cooperative v. Swidler, 228 F.Supp. 875 (1963).

Minnkota Power Cooperative v. Swidler, 228 F.Supp. 968 (1963).

Wisconsin v. Federal Power Commission, 373 U.S. 294 (1963).

Tennessee Gas Transmission Company v. Federal Power Commission, 340 F.2d 100 (1964).

Federal Power Commission v. Southern California Edison Company, 376 U.S. 205 (1964).

Tennessee Gas Transmission Company v. Federal Power Commission, 381 U.S. 950 (1965).

Salt River Project Agricultural Improvement and Power District v. FPC, 391 F.2d 470 (1968).

Permian Basin Area Rate Cases, 390 U.S. 747 (1968).

Index

This book was designed and typeset on a Macintosh computer system using QuarkXpress software. The text is set in Minion, and display type is set in Usherwood. This book was designed by Ellen Beeler, typeset by Bill Adams, and manufactured by Thomson-Shore, Inc.